America's New Downtowns

PUBLISHING FOR THE WORLD
125 Years
THE JOHNS HOPKINS UNIVERSITY PRESS

Creating the North American Landscape

Gregory Conniff
Edward K. Muller
David Schuyler
Consulting Editors

George F. Thompson
Series Founder and Director

Published in cooperation with the Center for American Places,
Santa Fe, New Mexico, and Harrisonburg, Virginia

America's New Downtowns

Revitalization or Reinvention?

Larry R. Ford

The Johns Hopkins University Press
Baltimore and London

The Johns Hopkins University Press
2715 North Charles Street
Baltimore, Maryland 21218-4363
www.press.jhu.edu

Library of Congress Cataloging-in-Publication Data
Ford, Larry.
    America's new downtowns : revitalization or reinvention? /
by Larry R. Ford.
        p.   cm. — (Creating the North American landscape)
    ISBN 0-8018-7163-8 (hardcover : alk. paper)
    1. Central business districts—United States—Case studies.
2. Urban renewal—United States—Case studies. I. Title. II. Series.
    HT167 .F675 2003
    307.3'416'0973—dc21

                                                    2002007933

A catalog record for this book is available from the British Library.

Contents

Preface

I have been visiting downtowns on a regular basis since I was a small child in Columbus, Ohio. I have always liked going downtown and never tire of exploring the nooks and crannies of the city center. In writing about and evaluating downtowns, I have been faced with several decisions regarding how to best depict them in the text. I have decided to emphasize descriptive writing, sketch maps, "cartoonish" models, and street-level photos over the use comprehensive plans, aerial photography, extensive statistical data, and detailed maps. By doing this, I hope to capture the essence of city places as the majority of visitors experience them rather than the professional perspective of the land use analyst or transportation planner. While the latter materials can be extremely useful, they do not always tell us what it is like to be there. Even as a geographer, I cannot put everything on maps that I would like to without making a cluttered mess of it. To find out where things are actually located, consult AAA, tourist materials, or the World Wide Web.

Obviously, the things that I like and dislike will differ from those of some readers. That is as it should be. My goal is to get people thinking about what makes a good downtown and why. Readers should obtain a variety of additional materials, including maps, data sets, and other types of photographs and begin to ponder the many as yet unexamined variables. My concern is that we are not paying enough attention to the spirit of the places that we call downtowns. Our old models and generalizations are out of date and we are only beginning to understand what we want our downtowns to be.

With these caveats in mind, please read the book, take some walks, and start to develop your own evaluation schemes. It really can be a lot of fun.

The Downtown Imperative and the Need for Comparative Studies

Downtowns are not like they used to be, but then, they probably never were. Like the death of Mark Twain, the death of the North American downtown has been greatly exaggerated. While it is undeniable that downtowns have suffered relative decline as a plethora of competing centers have blossomed in the suburbs and elsewhere, it is not clear that they have suffered an absolute decline. There is no doubt that the downtown is no longer *the* center of retailing and social life in most large metropolitan areas, but it is often the primary destination for tourists, conventioneers, high-level government and business decision-makers, and those seeking cultural experiences and some types of recreation. As a result, downtowns have become more specialized and monumental, in some ways emulating the ceremonial cities of ancient times. The "new" American downtown is quite different from the classic "central business district" that evolved in the last century, but we have yet to really define and monitor just exactly what it is.

Many have referred to the "renaissance" of downtowns, and the analogy may not be far from apt. Just as work-a-day medieval city centers were transformed into the settings for grand palaces and piazzas during the Renaissance, so too

have many American downtowns gone from being central business districts to places of monumental display and entertainment. The new downtowns are based as much on maximum visibility and appeal as on maximum accessibility. Impressive skylines, waterfront parks, festival markets, cultural centers, and professional sports venues are the things that attract attention. The promenade may be coming back as downtowns offer settings for lifestyles as much as places to earn a living.

This postmodern city of consumption and display has come under a great deal of criticism by academics who view it as a sort of inauthentic theme park.[1] I believe that these criticisms, while having some validity, are nevertheless exaggerated. While there certainly are places that have been "Disneyfied" into silly whimsicality, they normally occupy only a small part of even the most revitalized downtowns. Of much greater importance is the fact that downtowns are now beginning to offer opportunities for pleasant strolls along a waterfront, shady places to eat lunch, interesting streets to walk, and inviting settings to meet friends or have coffee. Such urban settings fulfill some very basic human needs, needs that were often ignored during, for example, the heydays of rapid industrialization and automobile-infatuated modernism.

While scholars should not hesitate to critique the follies and failures of American downtowns as they appear at the beginning of the new millennium, neither should we ignore the real progress that has been made in at least some cities. We need to find new ways to monitor change since, in many ways, downtowns have become something new and different and cannot easily be compared with those of the past.

The Downtown Imperative

It is of vital importance that we learn more about what downtowns are and how they can be made to work as central places for our metropolitan regions. We live in a throwaway society and the costs, both financial and social, are beginning to mount. While most American metropolitan populations have grown only modestly over the past two decades, they have expanded rapidly in area. Old industrial areas, residential neighborhoods, and even shopping centers have simply been abandoned as new nodes of activity have been developed. Our tax laws, zoning codes, and lending policies have all encouraged rapid write-offs and early abandonment. New strip malls are often built in a shoddy manner with little attention to design since it is expected that they will last only

a decade or two.[2] Even large shopping malls are being closed and discarded as newer versions open. Hundreds of thousands of houses sit vacant while millions of families struggle to find affordable housing.

There is already a vast literature focusing on the environmental and social costs of urban sprawl, and so I will not belabor the point here. My goal is to demonstrate that downtowns represent who we are and who we have been and they represent massive investments over many decades. They give identity, meaning, and character to our increasingly placeless urban regions. If we cannot learn to revitalize our downtowns and make them important places once again, what chance do we have of learning to save the shopping strips and housing tracts we have built in recent years? We will be forever trapped in an expensive and wasteful spiral. Downtowns are a vital component in the quest to understand how we can modify our policies and procedures so as to encourage the preservation of older existing urban landscapes and to minimize unaesthetic, sprawling development. Revitalizing downtowns and older neighborhoods constitutes an essential part of the emerging conservation ethic.

A more pragmatic and urgent aspect of the downtown imperative relates to the idea that, for many cities, downtown is the only game in town. Most large American cities have found it difficult or even impossible to annex new territory over the past fifty years, and so the new malls and "edge cities" are located in independent suburbs. As central-city industrial areas and commercial strips have declined, the city tax base has dwindled. In "strangled" cities such as Pittsburgh, Cleveland, Baltimore, and Providence, there are few viable nodes for economic development other than the downtown. The tax money needed to improve inner city neighborhoods has to come from someplace.

In addition, one can argue convincingly that most downtowns simply have a great deal more potential for the creation of the truly unique and exciting places that fickle American consumers seem to crave than most suburban alternatives. Successful downtowns offer far more variety than even the largest megamalls and theme parks, and the associated social and economic costs, while not nonexistent, are far lower. I like downtowns. Part of the reason that I am interested in monitoring what works and what does not in the centers of American cities is that I feel that downtowns are more interesting and more important than the great majority of competing nodes of activity. Downtowns, or at least good downtowns, exude aesthetic, functional, and social diversity to a far greater degree than do suburban office parks and edge cities.

Downtowns tell us who we are, where we have been, where we are going, and

at what speed. Because cities have been so concerned about keeping their centers vibrant over many decades, nearly every architectural and functional era is represented. The fact that buildings come in different ages, sizes, and degrees of repair means that there are a variety of rents and activities. A variety of people can also be found on the relatively user-friendly streets of downtowns. Samuel Johnson said, "When a man is tired of London, he is tired of life." I know that many people prefer the all-new, clean, sterile, automobile-dependent suburban developments precisely because they are devoid of the kind of diversity found downtown. However, when we completely lose touch with our own vibrant and exciting diversity, we may indeed be said to be "tired of life."

The need to have every trendy new architectural era and activity represented means that downtowns are often in a constant state of flux, and this has been true for a very long time. Long before urban renewal and interstate highway construction, downtown buildings were torn down and replaced on a regular basis. The current transformation from an industrial to a postindustrial / recreational landscape is but the latest chapter in the ever-changing downtown. These transformations have not, however, been without costs, both economic and social. The current revitalization of American downtowns, for example, has usually involved massive public expenditures and considerable government subsidies. This, however, is another story. Nearly everything in the modern world is related to government strategies and subsidies. Income tax write-offs of mortgage interest payments and taxing and zoning strategies that have encouraged the use of rural land for new development have certainly accelerated suburbanization, so it should come as no surprise that government policies have also played an important role downtown. The issue is perhaps most clearly illustrated by the case of Washington, D.C., where decisions made by the federal government removed many jobs from the central city even as federal programs spent vast sums of money to soften the impact of unemployment there. For example, federal money was used to build a subway even as federal decisions moved many jobs beyond its reach.

There have also been social costs involved in downtown revitalization. As downtowns have been rebuilt, older activities and older people have often suffered physical, or at least psychological, displacement and alienation. This too is another story and one that has often been told.[3] It also illustrates a problem that is commonplace throughout metropolitan areas, as rural districts and small towns are engulfed by suburban hyperdevelopment while disinvestment and ghettoization devastate many inner-city communities. In America, few

places escape change, and change is not always kind. On the other hand, it can be argued that cities must either change or die, and this is especially true of downtowns. Even cities that have strong historic preservation ordinances and architectural controls, such as Paris or Rome, have experienced massive social and economic change in recent decades.

Spatial Structure and the Problem of Downtown Data

These are important issues but they are not what this book is about. While I do not wish to ignore the underlying social and economic realities that have led to the (post)modern downtown, I feel that there are some other issues that need to be addressed first, some very basic geographic issues that largely have been ignored for the past forty years. My goal is to add to our understanding of just what the American downtown has been and what it is becoming as we enter a new century and a new millennium. I want to explore the architectural trends and spatial linkages that define the downtown today and to develop a useful and multilayered "model" of downtown structure. In other words, I feel that we need to describe downtown before we can critique it adequately. We have a burgeoning urban literature replete with words such as *downtown, inner cities,* and *suburban,* but these terms are much less geographically precise than they used to be, and the situation may be getting out of hand. For example, today both inner-city and suburban areas can be located just about anywhere in our vast metropolitan areas, and many cities have elite "suburban" neighborhoods that are much closer to the old city center than are some poor, minority communities. The use of these terms, therefore, may not only obfuscate but even delude those seeking meaningful understanding of phenomena and solutions to problems, especially when terms like *inner city* and *suburban* are used as explanations.

While downtown is much easier to delimit with some precision than many other types of districts, it is still not always easy to decide where its boundaries should be and how much they should change over time as some things are torn down and new projects appear. While many of the older cities of Europe have a consensus "historic core" defined by once or still existing city walls, the American downtown is a much fuzzier concept. Even the origin of the term contributes to the confusion. *Downtown* historically meant the central business district of a city and/or the physically lower district down by the harbor or river. In New York City, which probably gave us the term, the earliest business

district was not only down by the water but "down south" at the tip of Manhattan. Still, it is doubtful if "center city" or some other term would make things clearer.

A more precise definition of the vernacular downtown focuses on the idea of a "central business district" or CBD. During the 1950s and 1960s, many urban scholars sought to develop precise measures of the CBD based on such things as height indices, land uses, and land values. Much of this work is summarized in R. E. Murphy's 1972 book *The Central Business District: A Study in Urban Geography.*[4] In recent decades, however, interest in the topic has waned. We still talk about downtown and critique its alleged follies and failures, but we no longer define or describe it. The irony here is that over the past thirty years the American downtown has changed dramatically in nearly every way. The old definitions of a central business district are no longer very helpful, but nothing has emerged to replace them. Much of the urban literature thus critiques something for which there is no consensus definition. Even the census has been reluctant to take a stand on just what constitutes a downtown and what types of data would help us to understand its relative importance. Since there are no consensus definitions of what constitutes a downtown in American metropolitan areas, it is nearly impossible to obtain comprehensive data on downtown populations, employment, densities, open space, cultural attractions, or crime. Cities have wildly different notions of what the term *downtown* means.

Part of the problem is that the central business district, even if it can be defined precisely, constitutes only a small part of what we typically refer to today as downtown. Historically, a frame of support activities such as warehouses, factories, residential hotels, transportation facilities, grain elevators, and the like has surrounded the CBD or core. In recent decades, both the core and the frame have experienced massive changes as urban renewal and private redevelopment have occurred. While core office districts may not have experienced significant functional change, the frame has often been changed beyond recognition. Sports stadiums have replaced factories and hotels have replaced tenements. Parks, condominiums, and concert halls have been built where port facilities used to be, and vast convention centers occupy land that was once residential or industrial. Most downtowns have had a policy of horizontal expansion over the past two decades, actively colonizing new territories as former barriers such as rail yards and port facilities have decamped. What is downtown? Where is downtown? How big can a downtown be?

While we have ignored the changing spatial realities of the new downtowns, we have applied more precise definitions to major suburban developments. We now have more detailed information about North America's "edge cities" than we do about most American downtowns. Joel Garreau's 1991 book summarized the literature on suburban nodes and, in the process, created a sort of model of the phenomenon along with some fairly precise definitions of what an edge city is.[5] For example, according to Garreau, a proper edge city must have five million square feet of office space and six hundred thousand square feet of retailing and it must be "perceived as one place." While there are no procedures for analyzing scale and spatial linkages, we do not even possess these vague definitions for "downtown." Without good, spatial definitions of the districts we critique, all data becomes problematic. Boundaries and definitions are extremely important since the inclusion or exclusion of a few blocks can make for significant differences in, for example, employment and recreational opportunities or residential population.

The problem is that our data collection units rarely match our spatial realities. While both the academic and popular literature have focused on such things as downtowns, edge cities, and malls, data are collected by census tract, political unit, and private project. We may not have the data we need to make realistic comparisons between places. Edge cities and malls often have the advantage of being under single ownership or at least of being part of a planned project, and so boundaries can be delineated with some precision. Even here, however, there are problems of definition. Some edge cities, such as Midtown Atlanta, are really just extensions of downtown. Others, such as Alexandria, Virginia, are really old downtowns, which have been engulfed in a larger metropolitan area. Still others, such as Beverly Hills, are really just segments of a much larger spine of development.

The problems of definition are well illustrated by New York City. Where is downtown New York? Sometimes the term *downtown* is reserved for the one-half-square-mile area around Wall Street, as opposed to the larger one-square-mile area of "Midtown" near Rockefeller Center. One would be hard-pressed, however, to say that Midtown is not really the functional "downtown." Others would argue that nearly all of Manhattan below Central Park (or maybe 110th Street) is really downtown, thus making a central business district of about six square miles. To further complicate matters, a quick stroll across the Brooklyn Bridge leads to "downtown" Brooklyn, while a brief ferry ride can take you to downtown Hoboken, a stone's throw away. Boston provides yet another exam-

ple of definitional problems. The Back Bay developments around the Pruden-
tial Center are now almost certainly part of "downtown" even though it may
take longer to walk there from City Hall than it does to get to a new "suburban"
mall across the river in Cambridge. Because we have given so little attention to
defining spatial units, it is difficult to determine what is going on in a comparison of one downtown to another with regard to office space, retail sales, or tax
base. Every city has a downtown, but downtown data are not collected in any
consistent, systematic way. Rarely do census tracts even correspond to a meaningful definition of downtown.

I have no immediate solutions for these boundary and data collection problems, although for the future I would propose official definitions for downtowns so that we may monitor and guide trends and tendencies with greater
accuracy. In the short run I shall simply define the downtowns I examine as
best I can. In some cases, of course, the ranking I give to a particular downtown
will be greatly affected by where I choose to draw the line. This is as it should
be. Others can draw different boundaries and come to different conclusions.
I will simply examine a number of case study downtowns as I define them
and compare and contrast the often-conflicting trends affecting them. In the
process, I will suggest a descriptive model of the "typical" downtown at the new
millennium.

Rating Downtowns: The Need for Comparative Studies

Americans love rankings. It seems that with every new issue of several popular magazines there is a new rating system that promises to provide readers
with information about the best colleges, automobiles, diet programs, vaca-
tions, restaurants, or movies. In recent years, these rankings have been joined
by a variety of city rating systems that claim to tell us about the best cities in
the nation. These articles focus on such themes as "best place to live," "safest
cities," "best cities for jobs," "best college towns," "best retirement centers," or
some other measure of the quality of life in American urban areas.[6] While I
find many of these rankings to be interesting, nearly all of these rating systems
rely on aggregate data rather than observations of and experiences in real
places. Consequently, it is sometimes difficult to interpret the data. For example, since both cities and metropolitan areas vary so much in size, density, and
population, it is hard to really compare one to another. The city of Newark,
New Jersey, for example, occupies only twenty-four square miles while Okla-

homa City (with a similar population) covers over six hundred square miles. Such variables as data on housing characteristics are wildly disparate. Similarly, the San Bernardino metropolitan area is larger than many states, and so aggregate data is hard to pin down to an actual place.

My goal is to compare real places, that is, little hunks of territory that are small enough and distinctive enough to be experienced and remembered. Specifically, I compare and evaluate sixteen American downtowns as I define them. Most American downtowns occupy about one or two square miles, and so a visitor can easily walk from one end to the other in twenty minutes. In addition, most American downtowns contain monumental buildings and other features that can be viewed continuously throughout such a walk, thus providing visual linkages. With some effort, therefore, essential downtown characteristics can be memorized and internalized. The resulting images can then be used along with other types of data in the evaluation of downtowns as places.

As I developed my downtown ranking system, I moved back and forth between objective data, such as amount of office space, number of hotel rooms, and cultural venues offered, and the subjective experience of walking through the city. It follows, therefore, that some aspects of my ranking system may be quickly and widely accepted and noncontroversial while other aspects will be rightly perceived as somewhat individualistic and arbitrary. This is as it should be. I do not seek to have the final word on the subject, and the more debates I can inspire, the better. I have two major goals in carrying out this project: The first is to encourage scholars to look specifically at downtowns in new and hybrid ways, and the second is to encourage the careful observation of actual places in the development of various types of ranking systems.

Choosing the Downtowns

Much of the literature on the American downtown focuses on the very largest and most important cities. Architectural histories, for example, typically concentrate on the skyscrapers, department stores, apartment buildings, and mansions of New York City and Chicago. Similarly, discussions of landscape architecture and urban design usually emphasize the making of Central Park or the Mall in Washington, D.C. Boston, Philadelphia, and Los Angeles have also received a fair amount of attention. While this emphasis on major centers is hardly surprising—after all London, Paris, and Vienna have received the lion's share of attention in studies of Western European cities—it is not always

helpful for those interested in "typical" American places. This is even more the case for those interested in typical American downtowns. There is only one New York City and it is nothing like the American norm. The same can be said to varying degrees for Chicago, Los Angeles, San Francisco, Philadelphia, Washington, D.C., and Boston. It is difficult to generalize about these unique urban settings. Although these cities, and more specifically their downtowns, tell a great deal about American values, architecture, urban design, and social change, they cannot easily be emulated by smaller cities seeking useable role models. While some innovations in zoning, architecture, and planning can be copied, it is very difficult for smaller cities to hope to emulate places such as Midtown Manhattan or the Chicago Lakefront. I have therefore moved a step or two down the urban hierarchy for my case studies.

I decided to choose the main downtowns of selected metropolitan areas of roughly between one million and three million people. The rationale was that these are large enough to have the threshold populations for major attractions such as professional sports teams, symphony orchestras, art museums, influential newspapers and other media, and specialized shopping and entertainment. They are, according to central place theory, "civilizing" third-order centers. On the other hand, most of the actual downtowns are small and "normal" enough to be emulated by places further down the urban hierarchy.

Of course, even the simple task of selecting downtowns for metropolitan areas in this size category involved several complications not reflected in the simple census data. Some metropolitan areas, such as Orange County, California, the Patterson, Clifton, Passaic area of northern New Jersey, and the Greensboro, Winston-Salem, High Point region of North Carolina, have large populations but no single dominant downtown. Others, such as Newark, New Jersey, or Oakland, California, are clearly small subcenters of much larger consolidated metropolitan areas. In still other cases, the PMSA (primary metropolitan statistical area) figures were significantly different from those for a CMSA (consolidated metropolitan area). With some exceptions, I used the CMSA figure since that tends to better describe threshold population.

Still, there were some fifty possible downtowns that could be used as case studies given my guidelines. As I sought to pare the list down to a manageable number, I tried to keep a wide variety of types of places that would allow me to talk about a full range of issues and characteristics. For example, I wanted to look at downtowns for metropolitan areas of varying sizes within my parameters. I also wanted to look at downtowns in various, although not necessarily

all, regions of the country. Finally, I wanted to include downtowns that epitomized different problems and characteristics from aging industrial settings in the Midwest to booming recreation centers in the Sun Belt. I wanted cities with interesting and challenging physical sites as well as those with few physical obstacles or constraints. I wanted diverse central places and downtowns with more specialized employment.

Above all, I wanted to look at downtowns that I knew pretty well and had known over a period of several decades. Nearly all of the sixteen downtowns that I finally chose to examine are places that I have visited many times over a forty-year period. I have some sense of how they got to be what they are today. I visited these downtowns and interviewed planners and academics as many times as I could during the period 1995–2000. I know more about some places than others and there is no doubt much that I have missed. Still, I have a pretty good sense of what these downtowns are all about. Once more I wish to emphasize that I am only looking at downtowns and not at cities or metropolitan areas. I am using the following metropolitan populations only as selection guidelines, and my purpose is to indicate the potential of particular downtowns as central places for a region.

The Downtown Case Studies (Metropolitan and City Populations)

1. Atlanta. 3.2 million, 404,000
2. Seattle. 3.2 million, 537,000
3. Cleveland. 2.9 million, 496,000
4. Minneapolis. 2.7 million, 352,000
5. San Diego. 2.6 million, 1,220,000
6. Phoenix. 2.5 million, 1,200,000
7. St. Louis. 2.5 million, 340,000
8. Baltimore. 2.5 million,* 645,000 (*Baltimore PMSA)
9. Pittsburgh. 2.4 million, 340,000
10. Denver. 2.2 million, 500,000
11. Portland. 2.0 million, 504,000
12. Indianapolis. 1.5 million, 741,000
13. San Antonio. 1.4 million, 1,114,000
14. Columbus. 1.4 million, 670,000
15. Charlotte. 1.3 million, 505,000
16. Providence. 1.1 million, 151,000

The above list is not a perfect representation of the American urban scene but it does include a wide variety of types of downtowns. The downtowns of Cleveland, Baltimore, and Pittsburgh, for example, were once surrounded by heavy industry, railroad yards, and port facilities. For the most part, downtown revitalization revolved around the reuse of these industrial landscapes. On the other hand, the downtowns of Phoenix, Charlotte, and San Diego are much newer places with relatively little core industry. Building a downtown usually means rebuilding low-density residential areas. The downtowns of Columbus, Indianapolis, and Atlanta lie somewhere between the two extremes. Similarly, Seattle, Portland, and Pittsburgh have challenging physical sites while San Antonio, Indianapolis, and Columbus have few physical obstacles to contend with. Some downtowns, such as those of Phoenix and Seattle, serve as the centers of booming metropolitan areas while others, such as downtown Cleveland or Pittsburgh, are important places in relatively slow-growing or even declining regions.

I have also sought to include case studies from most of the regions of the country, although I have not achieved an even sampling. I have lived in the Midwest and West Coast but not the South, and so I am more familiar with cities in the former regions. In addition, most southern metropolises, with a few exceptions like Atlanta, are only now approaching the one million mark, with smaller cities predominating. The Northeast also presents some problems in that the megalopolis dominates the urban hierarchy to the point of minimizing the importance of the few third-order cities that exist. The Midwest, on the other hand, has too many appropriate centers and so, with reluctance, I dropped such places as Cincinnati, Milwaukee, and Kansas City.

Western central places have the advantage of being very dominant in their respective regions. Denver and Phoenix have little nearby competition for major attractions compared to, say Cleveland or Baltimore. The list contains a great deal of variety since Providence, Pittsburgh, Phoenix, and San Diego are very different kinds of places with very different downtown characteristics. Still, many trends, successes, and failures are shared.

Identifying the Variables

In order to systematically compare and evaluate the spatial characteristics of these sixteen American downtowns at the millennium, I identified ten variables for usage in a detailed analysis of each place. While the downtowns are

large and vibrant enough to have all the necessary components for the "new" American downtown and may serve as possible role models for smaller cities, their attributes are unevenly distributed. Some cities may rate very highly on one variable but fall down on another. The methodology is simple and straight-forward. Using both objective data and subjective experiences, I rank each downtown on the ten variables discussed below. Every downtown receives ten individual "grades" and an overall composite score.

Downtown Variables

1. Physical site: the display/use of that site including waterfront, topog-raphy, views, recreation potential, sense of place and history; access.
2. Street morphology: reinforcing site with streets; great streets and cozy streets, scale, flows, gaps, access, sense of center, size of blocks, alleys.
3. Civic space: public architecture and parks, central squares and plazas, public art and other embellishments, cultural facilities such as li-braries and colleges.
4. Office/skyline: the amount of space and downtown employment; the character, variety, ambience, age, and architecture of office buildings; number of headquarters, landmarks, and the sense of a financial dis-trict.
5. Retail-anchors—the CBD as a shopping destination: inside (malls) and outside (street-level) shopping, variety of markets, socioeco-nomic range, basic needs and specialty shops, local as well as national stores.
6. Hotels/convention facilities: visitors and the twenty-four-hour city, convention centers and linkages to them, support districts, number of hotel rooms in the core and nearby, ability to attract major shows and visitors; downtown tourism.
7. Major attractions: big draws for big crowds, sports venues, museums, art galleries, theater districts, aquariums, special event sites, the sense of spectacle and celebration.
8. Historic districts/support zones: human-scale "places," walkabouts, sense of heritage and history, places for the arts and petty entrepre-neurs, light support industry, functional as well as architectural preservation, ethnic heritage; intra-downtown district identity.
9. Residential activity and variety: condos and apartments in new and

renovated buildings in the core and connections to nearby residential neighborhoods and support districts (schools, shops, gas stations); a variety of luxury condos, market-rate apartments and lofts (recycled space), and affordable housing such as single-room-occupancy hotels (SROs).

10. Transit options: public transit and negative auto impacts; ease of movement, light rail, bus malls, and bike trails; the impact of auto-related features such as freeways, garage ramps, massive parking structures, and bleak throughways; sidewalks and walkability.

Adopting a Stance

Since much of this book is about subjective interpretations, I need to be explicit and up front about the kinds of things I like and don't like. I am adopting a stance that I feel is a combination of romantic/utopian notions of a perfect city with a pragmatic recognition that there are certain kinds of functions and spatial characteristics that modern downtowns must have in order to be viable and competitive.

The romantic/utopian stance suggests that I favor fairly traditional, compact downtowns in which, for example, attractive buildings enclose lively plazas. I dislike sprawling, empty downtowns with wide, traffic-choked highways. I like places where things are close together and people can easily walk to a variety of attractions on interesting sidewalks with lots of street-level shops. I dislike massive fortress buildings with blank walls facing the street. I like places that have interesting physical settings such as riverfronts and that make use of them effectively. I like places that have preserved at least part of their histories and display a vivid sense of place. I dislike bland, homogenized settings that seem placeless and interchangeable. In all of these things and more, I adopt a fairly traditional, pro-urban view of what makes a good downtown. On the other hand, I feel that something more is needed in order to avoid a clichéd checklist of admirable downtown characteristics.

On the pragmatic side, I recognize that, like them or not, downtowns need certain things that do not always reflect and may even contradict the urban design policies described above. For example, downtowns must have large convention centers and parking garages even though they often include at least a few massive, blank walls and sidewalk-disrupting entryways. American downtowns, unlike those of Europe, must also have a relatively legible street system

with few confusing and awkward intersections since they are dependent on ve-
hicular traffic as well as pedestrian flows. I also recognize that, while they are
sometimes controversial, skyscraper office buildings and hotels, sports arenas,
and waterfront festival marketplaces are some of the things that downtowns
can do well. As downtowns and suburban nodes become more specialized,
downtowns must build on their strengths even when some aspects of success
may be problematic. I realize, for example, that downtown baseball parks and
performing arts centers are often financial boondoggles and that tax monies
are not always wisely spent in the pursuit of grandeur, but still I generally ap-
plaud downtown efforts to become the "best game in town." On the pragmatic
side, if average residents, investors, government regulators, and tourists like the
place, I am willing to concede that the downtown is successful. In spite of my
romantic and utopian concerns, I am willing to be flexible. I try not to focus
too much on intangibles such as "authenticity" or traditional renaissance notions of aesthetics.

I am also focusing on place at the expense of process. I am looking almost exclusively at spatial and landscape characteristics. I pay little attention to such very
important attributes as effective leadership, local business and governmental
culture, role in the changing national economy, or historical accident in the making of the places I critique. I realize that places are created by people and that to
fully understand how places came to be, it is necessary to understand the roles
of important and diverse "movers and shakers" who made pivotal decisions that
shaped downtown growth in particular urban contexts. Without people like
Robert Moses, Richard Lugar, Andrew Carnegie, the Mellon family, and a host
of other major players, America's downtowns would be very different places.[7]
But this is another story and one that has been at least partially told in the existing literature. I am satisfied to know what the downtowns are like today.

A very important thing to keep in mind is that I am looking only at downtowns—not cities or metropolitan areas. In some cases, downtowns that I rank
very highly are located in cities with major problems. In other cases, poor
downtowns form the centers of booming metropolises. Controversies arise, of
course, such as when money is spent to revitalize downtown while city neighborhoods are decaying. This too is another story. I am evaluating downtowns
as places and will not deal with issues surrounding downtown versus neighborhood investment strategies. Still, I am adopting the stance that without the
downtown as an economic engine, most declining cities would be even worse
off than they are today.

For reasons I outlined in the introduction, looking only at downtown is more difficult than it sounds. What and where is downtown? What nearby neighborhoods should be included in downtown support districts? What major parks and physical amenities are close enough to downtown to be included in the evaluation? These are all necessarily subjective decisions.

A Guide to the Following Chapters

Building a Context

The book is divided into an introduction and six chapters. The introduction and the first two chapters are aimed at building a context for the comparative studies that come later. In chapter 1, "The American Downtown: The Myth of a Golden Age," I discuss the popular image of downtowns and their changes over time. It is hard to find neutral literature on the American downtown. Most contains either glowing, promotional accounts of success stories in one or more cities, or pessimistic studies of the diminishing role of downtowns and / or the socioeconomic problems that have resulted from both central-city decline and attempts at revitalization. Much of the academic literature falls in the latter category, since we are taught to be, above all, critical and skeptical. Still, our critiques might be more useful if anchored to better-detailed information on what downtowns have been like historically, what they are like today, and how they vary one from another. For example, many commentaries imply that downtowns have plummeted in stature from a once golden age. It is more likely that some things have changed for the better while others have changed for the worse and that it may be hard to find a time when the consensus was that downtowns were wonderful places. In a nutshell, I critique the critiques.

In chapter 2, "The Evolution of the American Downtown: 1850–2000," I briefly trace the evolution of the American downtown from its inception to current projects. Here, I discuss the origins of downtown in the context of the changing scale of economic activity, changing transportation technologies, and the development of special-purpose architecture. What we have come to know as downtown or the central business district is really a fairly recent phenomenon, coming into its own only toward the end of the nineteenth century. Before the advent of skyscraper office towers, large department stores, luxury hotels, apartment buildings, train stations, and a variety of industrial buildings, downtowns as we know them did not really exist.

The classic, highly centralized downtown evolved in the early decades of the twentieth century. It was no sooner fully developed than automobile-related suburban developments began to challenge its dominant role in the life of the city. As a result, the second half of the twentieth century has witnessed a variety of strategies and planning efforts aimed at trying to keep downtowns alive and healthy. Some of these strategies have worked better than others. This "defensive" era can be divided into a modern (1950–1975) and a postmodern (1975–2000) period, each having different and sometimes conflicting goals, policies, and priorities. The first period, for example, emphasized modernistic architecture, superblocks, freeway inner belts, and large-scale slum clearance while the latter has focused on historic preservation, festival marketplaces, upscale housing, mass transit, and the development of lively streetscapes. Of course, the actual time frame varies from city to city, but for the most part the change in orientation was universal and clear cut. It should be evident by the end of this chapter that the new millennium will bring us both change and continuity.

Discussing the Variables and Ranking the Downtowns

Chapters 3, 4, and 5 deal with the ten variables that I have chosen to use in the evaluation of downtowns. The variables are grouped into "packages" with three in chapter 3, three in chapter 4, and four in chapter 5. For example, in chapter 3, "The Downtown Stage: Physical Site, Street Morphology, and Civic Space," I begin a discussion of the specific downtown evaluation process with a look at the three variables listed in the chapter title. I look at how downtowns have been both constrained and inspired by their physical sites and how street systems and public spaces have further shaped the urban core. The basic theme is that most downtowns are located in relatively interesting and challenging physical settings. While most large suburban developers have sought out relatively flat, easy-to-develop sites, downtowns usually have had to deal with some combination of harbors, rivers, hills, peninsulas, and valleys. They were located originally at the confluence of early transportation routes, which normally meant building, for example, where rivers met the sea or where mountain passes opened onto the plains. Thus, downtowns are often far more interesting places than competing edge cities, if only from the standpoint of physical geography. In cities such as Pittsburgh, Seattle, Portland, and Baltimore, much of downtown identity and sense of place is inextricably intertwined with the physical setting. Indeed, the need to incorporate waterfronts, scenic views, and

parklike settings into the downtown image ranks high on the agendas of many planners.

The basic spatial arrangement or morphology of most downtowns is often related to these topographical features, but there are other factors as well. The organization of the central district is very much dependent on street layout and the connections between adjacent neighborhoods and functions. Some cities, for example, have one large grid with wide streets and big blocks throughout downtown, while others have several competing grids with varying block sizes and street widths. In some cities, access to adjacent areas is easy in every direction, while in others rail yards and major highways break up the downtown into separate fragments. Some downtowns have long blocks with alleys, while others have short blocks with none. Some downtowns have monumental squares and parks, while others are devoid of any significant open space. The morphology of the downtown, in combination with physical features, provides the stage upon which people act out the life of the city.

Civic architecture and public space, such as central squares, grand boulevards, parks, and civic buildings, have also shaped the downtown setting. American downtowns often lack the grand public spaces and strong sense of center common to the more traditional cities of Europe and Latin America. In other parts of the world, a palace, city hall, cathedral, and grand hotel were often purposefully grouped around an impressive "plaza major." There was no doubt where the center of the city was located. Offices and retail activities were usually located on nearby, less important streets, supporting but not intruding into the civic center. Grand, tree-lined boulevards of sometimes monumental proportions also often defined the center in such cities. In America, civic structures are usually both less grand and less centrally located than they are in, for example, European or Latin American cities. State capitols, for example, may be monumental in size but are often located well away from the consensus center of town. In some cities, civic space does come close to defining the core of the CBD, while in others it is all but invisible. Educational institutions also fit under this heading. Universities, community colleges, art, law, and business schools, and teaching hospitals are sometimes major employers in a downtown area and help to give it character.

How civic spaces are used is also an important variable in measuring the importance of a downtown. Many downtowns have once again become important for celebrating a sense of time in the city as well as a sense of place. Parades, street fairs, political rallies, fireworks, seasonal decorations, music con-

certs, marathons, and art shows have evolved to showcase the public spaces of revitalized downtowns. Sometimes it seems that the new downtown is as much a place for spectacle as for traditional work.

Chapter 4, "Traditional Downtown Functions: Offices, Retailing, Hotels, and Convention Centers" suggests that most American downtowns are more often dominated by these four essential economic activities than by political or religious structures. The biggest and most expensive buildings usually define the center of the city and draw people to them. Like it or not, without large amounts of quality office space, often in landmark towers, significant upscale retail, luxury hotels and convention facilities, and major cultural and sports attractions, American downtowns are not likely to be seen as competitive.

The "big gorilla" that drives most large American downtowns is office space. Beginning around the turn of the twentieth century with the advent of the skyscraper office building, office space became the most important land use at and around the emerging PLVI or peak land value intersection. As corporations built bigger and more opulent landmark structures, the city skyline became part and parcel of the image of the American downtown. Over time, department stores joined in the race for biggest and best with Macy's department store in New York City exceeding two million square feet of space. Towering hotels and, later, sprawling convention centers gave prestige to the urban core. One result of the competition for more prestigious facilities has been geographic change over time. As newer and bigger buildings were constructed, the core of the downtown often moved, leaving behind less competitive structures. American downtowns vary greatly in the amount of space given over to core activities and in spatial organization, density, and types of linkages. Some downtowns are very compact with a few well-connected large buildings while others have numerous smaller and somewhat separate districts. Some seem congested while others seem more open. What are the trends in the location of these downtown components?

Chapter 5, "Downtown Expands: Major Attractions, Historic Districts, Residential Neighborhoods, and Transportation Innovations," focuses more on the fringe or "frame" of downtown than on its core. Many of the activities found in these areas were not even considered to be appropriate downtown land uses a few decades ago. Examples include residential areas, historic districts, industrial support zones, and nearby commercial "midtown" neighborhoods. Such districts often create their own "world" within the downtown and so transit linkages such as light rail or bike paths are needed to tie them to-

gether. These peripheral areas provide settings for entertainment and recreation in the form of small cafes, galleries, used book stores, and housing in the form of lofts, apartments above shops, and luxury condominiums, especially if they are part of an officially designated historic district. They are also often the home of controversial stadiums and arenas, which can help provide customers for restaurants and cafes. These new types of spaces loom large in the image of what a "big-league city" should be. But there are controversies. One is the issue of homeless shelters and associated social service agencies. Where can they be located now that skid row has become "olde towne"?

As city centers have deindustrialized, a tremendous opportunity has arisen to "recapture" space once used for manufacturing and warehousing. On the other hand, planners in some cities have argued that this process has gone too far and have taken steps to ensure that urban renewal policies and escalating land values do not force industry out. While heavy, noxious industries are increasingly absent from downtowns, there is still a need for lighter industrial support facilities. These might include such things as distribution centers, electrical and janitorial services, computer repair operations, and some remnant port activities. These functions make for a greater variety of downtown employment and provide threshold populations for additional eating, drinking, and shopping venues.

Housing is an important variable too. One can argue that, until recently, living downtown has not been considered to be quite normal in the American city. By the early twentieth century, reformers were attacking the residential hotel as an inappropriate place to live, and even large apartment buildings were viewed with a great deal of alarm as potential sites for "indecent propinquities." In addition, much of the downtown population lived in outdated, remnant shanties left over from the years of initial settlement, usually the first structures slated for urban renewal. On the other hand, in recent years nearly every city has adopted a policy of encouraging people to live downtown.

Perhaps inspired by the vibrancy of cities in other parts of the world, downtown planners began to seek a residential population that would support a twenty-four-hour city. Here the problem of downtown definition is especially acute since *by definition* residential areas were excluded from most early studies of the CBD. Are neighborhoods such as Beacon Hill in Boston or Nob Hill in San Francisco considered to be "downtown"? Perhaps we need new definitions of the greater downtown region in order to measure the importance of downtown living. One of the keys to a successful downtown, as illustrated by

examples from New York, Boston, and San Francisco, is that there is a seamless connection between the downtown core and relatively stable, prosperous neighborhoods. One can walk from Rockefeller Center in Manhattan to the elite residential streets of the Upper East Side without encountering vast barriers and unfriendly obstacles. Many American downtowns, on the other hand, are surrounded by heavy industry, railroad yards, abandoned buildings, seas of parking, freeway interchanges, or skid rows. Such downtowns are often physically and psychologically disconnected from the rest of the urban area. Of course, the transportation infrastructure plays an important role as well. Cities with good mass transit options such as subway or light rail systems can usually support high-density, consensus nodes better than those without.

Conclusion

Finally, in chapter 6, "Ranking Downtowns: Toward a Model of Spatial Organization," I summarize the strengths and weaknesses of the various downtown case studies and give each a final score. In so doing, I both differentiate the downtowns and identify some common characteristics and trends. I suggest some possible models of downtown structure building upon data from the sixteen cities and speculate on what downtowns will become as the new century unfolds. What is the typical downtown today and what changes are afoot as we enter the next millennium? I also ponder the convergence and divergence of central cities as the global economy increasingly fosters both homogeneity and interesting hybrids.

The book is amply illustrated with both photographs and maps so that the reader can use them in order to arrive at individual speculations about future urban landscapes and spatial arrangements. I will make many observations, but the goal is to foster discussion and inquiry, not to proclaim the truth.

The American Downtown

The Myth of a Golden Age

Much of the writing on American downtowns has been quite critical, often to the point of being contradictory. Downtowns have been described, for example, as both too small-scale and obsolete and too massive and futuristic. They have been said to be both too empty and too congested, too expensive and too run-down, too large in area and too compact, too exclusive and too diverse, too bleak and too whimsical, and either overly controlled by an elite power structure or too devoid of meaningful leadership.[1] The contradictory nature of downtown critiques results from two sources of confusion, geography and history. First, in spite of seemingly homogenizing trends, big-city downtowns are really becoming more different from one another than they have been at any time in the twentieth century. More than ever, place matters. We need to be more specific about the particular downtowns we are describing when we enumerate their problems and prospects. It is possible that far too many of our generalizations about cities come from New York City and Chicago. Most American cities, for example, have never had tenement districts, massive port facilities, or huge ethnic neighborhoods anything like those common to a bigger metropolis. For the purposes of this work, I am focusing on the downtowns

of metropolitan areas with between one and three million people. Both smaller and larger cities may have different problems and different histories.

History matters too. When both academics and the popular press criticize downtowns, they often assume that things were once much better, at least with regard to the category under discussion. There is the assumption that downtowns have declined from some golden age when current problems were less pronounced or even completely absent. I will deal with the first source of confusion later on when I examine particular case studies. In this chapter, I focus on the second source of confusion—the myth of decline from a golden age.

In his book *The Idea of Decline in Western History*, Arthur Herman points out that the perception of decline from some golden age is not only deeply embedded in the Western tradition, but is a part of nearly every culture's story of origin.[2] He suggests that intellectuals have been predicting the imminent collapse of Western civilization for more than 150 years but that the pace of the alleged decline seems to be quickening. Today, we not only romanticize preindustrial societies when people lived in harmony with nature, but also Fordist industrial societies when blue-collar workers had job security and good pay. In much the same way, we have focused a great deal of attention on the topic of urban decline and the crisis of the modern American city. Downtowns, in particular, are seen as either in retreat from an earlier golden age or in the midst of desperate attempts at revitalization. The idea of decline requires myths as well as facts.

There is one critique of downtown, however, that is both straightforward and undeniable, and that is that downtowns are relatively less important than they used to be. Big-city downtowns rarely have even 10 percent of the retail trade in a metropolitan area, and many have less than half of the supply of primary office space. Industrial and warehousing activities have rapidly decamped from the central city and so, as the center of employment and business, the one-and-only "central business district" idea may indeed be passé. But is this a bad thing? The idea of having heavy industry and massive rail yards and port facilities in and around downtown was probably a bad one in the first place, and much of the decline in these activities is due to purposeful attempts to rid the central city of inappropriate land uses. By eliminating many kinds of noxious and space-extensive functions, American downtowns may be simply returning to an idealized preindustrial pattern epitomized by many of the more attractive cities of Europe. When industrial pollutants led to a fire on the Cuyahoga River in downtown Cleveland during the 1960s, it became clear that "Paris was right": the central city should be for people rather than for heavy industry.

On the other hand, the relative decline in "clean" activities such as primary office space, upscale shopping, dining, and cultural activities in downtowns has been perceived as more of a problem. Even here, though, the data can be misleading. Given the explosion in retail, service, and restaurant space over the past few decades, it would be a disaster if downtowns were the home of even a quarter of the discount stores, automobile dealers, home improvement emporia, and neighborhood service centers that are currently found throughout metropolitan areas. As urban areas have expanded to engulf the downtowns of smaller cities and as new kinds of space-extensive and auto-dependent retailing have been invented, it only makes sense that the CBD no longer includes a very high percentage of the total economic activity in the metropolitan region. Many types of office space need not be downtown either. The boom in semiskilled "back office" employment has rightly led to the suburbanization of many clerical functions. The point here is not that downtowns are relatively less important in terms of most, if not all, economic activities but that this is a meaningless finding. It would be absolutely inconceivable for it to be otherwise. It is not now nor was it ever a good idea for people to have to take mass transit downtown in order to buy a bag of fertilizer or a lawn mower.

The only meaningful question is, has the downtown experienced an absolute decline in the activities that should logically be concentrated there? If a downtown has suffered significant absolute losses in prime office space, high-end retailing, cultural venues, entertainment, and public space, then there is cause for alarm. For many everyday activities, such as grocery stores and barber shops, on the other hand, a downtown need only have enough to serve its resident population.

Once we accept the fact that downtowns have gotten to be relatively less important than they were at one time, we can return to the topic of the myth of a golden age and examine some of the other oft-mentioned criticisms of downtowns. There are a number of what I shall call myths and assumptions that are embedded in many of the critiques of downtowns. In *The Way We Never Were*, Stephanie Coontz relates the ways in which we have romanticized and mythologized American family life in the past.[3] To a very real degree, we have done the same thing for the American downtown. We have created a set of mythical images of the wonderful downtowns of the past. Each myth provides a kind of romantic haze that serves as the basis for critiques of the current scene. This is not to say, however, that real problems do not exist and that downtowns are

functioning perfectly as central places for their urban areas. Rather, critiques of downtowns are most often a fuzzy mix of myth and reality.

So, Just When Was the Golden Era?

If we are to organize our critiques of downtowns around the idea that they have declined from some golden era(s) in the past and that they are no longer as good or as important as they once were, we must specify just when, exactly, those ideal periods were. It may be that every period has had both good and bad characteristics and that nostalgia for the past is a necessary by-product of constant change. We may be simply lumping together all the good memories from various past eras and comparing the whole assemblage to our current condition. We need to look more carefully at conditions in particular decades in order to make valid comparisons.

Generally speaking, when reverence for the traditional American downtown comes up, the focus tends to be on some part of the period between roughly the late 1880s and the early 1960s. Before 1880, most American downtowns were not yet fully formed; indeed, many southern and western cities had barely been founded. Central areas were roughly thrown together with smallish, wooden, multipurpose buildings that burned down periodically. The infrastructures were minimal and basic services such as paved streets, reliable water supplies, street-lights, and mass transit were not always in place. The architectural and functional components of downtown were not fully developed either. Office buildings, department stores, apartment buildings, shopping arcades, luxury hotels, theaters, and monumental public buildings came into their own during the last decade of the nineteenth century and the first two decades of the twentieth century. In addition, the City Beautiful Movement that began in the 1890s led to the creation of a number of important civic spaces. Before then, downtowns were still in a state of becoming and most were pretty ragged. Perhaps the golden age did not begin until the 1920s, marred only by Al Capone and strict racial segregation.

By the late 1960s, most American downtowns had clearly lost some of their luster. Suburbanization, urban renewal, highway construction, lack of investment during the Depression, redlining, riots, and the passage of time had all taken a toll. Government plans and policies as well as private investment played an important role in the changing image and reality of downtown. Just as governments had encouraged the City Beautiful Movement at the beginning of the century, they seemed bent on creating suburban utopias by midcentury.

But even the eighty-year period between 1880 and 1960 was full of ups and downs. The 1920s were years of relative prosperity and innovation but also of coal smoke, prohibition, gang wars, and Ku Klux Klan parades. The Depression years of the 1930s are remembered for grand musicals at the new movie palaces but also for bread lines and disinvestment. The war years of the 1940s were both a time of trauma and a time of camaraderie. Rationing and shortages made for a sense of mutual sacrifice and community. Many of the most famous photos of happy downtown crowds featured victory celebrations in 1945, but in reality, such crowds were a rarity. In spite of these various shortcomings, we can probably identify some periods that have been widely portrayed as the leading contenders for golden era status.

Three Possibilities for Golden Era Status: 1910, 1928, and 1950

There would seem to be three possible time periods that might qualify as epitomizing some notion of a golden era for American downtowns: the City Beautiful Era, the Roaring Twenties, and the Booming Fifties. Rather than identify exact temporal parameters for each of these eras, I am suggesting specific years to represent them. The first of these, 1910, represents the height of City Beautiful planning and urban embellishment. Inspired by the Chicago World's Fair in 1893, cities all around the nation began to design and build grand, monumental civic centers complete with new city halls, libraries, opera houses, fountains, grassy malls, and grand boulevards. While the impact of the City Beautiful Movement was uneven, affecting some downtowns far more than others, nearly every American city benefited from new notions of good planning and civic identity. The "thrown-together" downtowns of the nineteenth century began to take on European-inspired Baroque embellishments. Famous examples include the civic centers of New York, San Francisco, Denver, and Cleveland, along with the reorganization of the Mall in Washington, D.C.

In addition to grand planning, this era featured a wide variety of new and exciting urban features such as electric streetcars, electric lights, early skyscrapers complete with speedy elevators, palatial vaudeville houses, grand apartment buildings, and a number of other novel attractions. Obviously not all of these features sprang onto the scene in 1910, but this prosperous, prewar period still represents a time when downtowns were innovative attractions. It was also a time when their sense of the civic role was quite strong.

A second possible golden year is 1928, the last full year of the Roaring Twen-

ties. A building boom was under way that led to some of the tallest towers, biggest movie houses, most luxurious apartment buildings, and largest department stores ever constructed in American cities. Transportation options were greater than ever with streetcars, motor buses, commuter trains, automobiles, and, in a few cities, elevated trains and subways offering ways in and out of the city. The construction of green parkways helped to facilitate pleasant commutes by automobile. Talking pictures brought people into the city for entertainment, and large numbers of female clerical workers made downtowns more gender-balanced. The combination of skyscrapers, huge electric signs, and prohibition made downtown a somewhat "racy" place where flappers did the Charleston in the midst of illicit refreshment. At least these are the images that have persisted. [4]

Perhaps the period just before and after 1950 constitutes the most relevant golden age for those who still remember when downtown was king and the return of prosperity and the end of rationing meant that there was money to spend. During the early and prosperous fifties, people were shopping and offices were humming. While suburban housing tracts were beginning to appear, nearly everything of any economic importance was still concentrated downtown. While the total percentage of retail downtown had already declined, much of this was due to the growth of neighborhood-level stores such as supermarkets and auto dealers. The big department stores and fancy restaurants were still downtown.

But these potential "golden eras" epitomized downtown problems as well as successes. For example, though downtowns seemed lively during the early 1950s (the only golden age that is still widely remembered), after years of deprivation and rationing, they also seemed tired and a bit worn out. Nothing much was built or even remodeled between 1930 and 1945 and a lot of shops and offices were looking a bit past their primes. Proclaiming that things were better then and that we are just now experiencing a "renaissance" aimed at restoring the glories of the past may be a bit far-fetched. Indeed, one might ask, if downtowns were really so lovely and lively during the 1950s, why was everyone from government agencies to private investors so eager to either tear everything down or cover it up with "modern" plastic or aluminum? Streetcar systems were destroyed along with nearly all of San Francisco's cable cars. "Out with the old and in with the new" once again became the motto of the day; only this time the game had changed. The combination of suburban competition and misguided modernist planning served to make many downtowns seem less inviting.

By the late 1950s, downtowns were also kind of tired in a cultural sense. There was little novelty involved in going downtown. Old buildings and old formalities remained even as rock and roll and drive-ins provided hints of the cultural changes ahead. In order to thrive, downtowns would have to change to meet the needs and demands of the coming era, especially in the face of competition from new suburban malls. Something had to be done, but what? Most planners and designers either had no clear idea of what the downtown should eventually be or they were enamored of the prospect of extreme modernist images of the city as skyscrapers in a park, all but devoid of life at street level. The immediate goal was to get federal money to clear slums and build highways.[5]

While the changes have sometimes been both too slow and painful and too rapid and disruptive, many downtowns have gradually been transformed over the past four decades into something better suited to their new roles in the metropolitan economy. Glaring mistakes have been made, but some important lessons have been learned.

When it comes to the issue of downtown revitalization, it may be a matter of seeing the glass as half full or half empty. While many downtowns may not be the dominant centers of activity they once were, things do seem to have turned a corner. Many American downtowns experienced a low point between the late 1960s and early 1980s. A variety of government and private-sector policies and procedures resulted in the destruction of vast numbers of downtown buildings and neighborhoods while encouraging and subsidizing wasteful urban sprawl. During the 1970s, many older downtowns were nearly obliterated and several cities, New York and Cleveland among them, went bankrupt. Factories and department stores closed and office buildings, when they were built at all, were designed to be fortresses complete with in-house parking and cafeterias. The traditional downtown was on its way out and, for a while, there was little consensus about how, or even if, it should be replaced. Indeed, our use of the term *renaissance* to describe more recent developments should be viewed in the context of the ups and downs of the past few decades.

Along the way, however, a consensus did begin to build. Important books were published, protests were organized, new types of landmark projects were developed, and old mistakes were recognized. Jane Jacobs's *The Death and Life of Great American Cities* (1961) and several later books helped to change the way people thought about good cities.[6] The historic preservation and adaptive reuse movements were jump-started by such projects as Ghirardelli Square in San Francisco (1964) and Quincy Market in Boston (1976). The residents of

SoHo in New York City managed to stop a cross-town freeway and new life came to lower Manhattan. In recent years, freeways have been torn down or put underground in Portland, San Francisco, and Boston, and some say that freeway removal will be an important part of the public works agenda in many cities in the twenty-first century.

The boom of the 1980s brought a great deal of investment in American downtowns as well as in suburban malls and office parks. While the boom brought a glut of office and retail space to nearly every locale, it did demonstrate that, with money, downtowns could be made fun and attractive once again.

Downtown Problems and Downtown Myths

When downtown problems are considered in the context of a decline from an alleged golden age, it is wise to examine both long-term continuity and change. While there may be some very new and serious problems facing American downtowns, it seems that many have been exaggerated by comparing the modern scene with a mythological and romanticized past. As I review a list of twelve major downtown "problems," I will attempt to explore some of the myths that could serve to make these problems seem newer and perhaps more serious than they really are. In the remainder of this section, I list the problems with the related myths in parentheses.

Problem (Myth) One: Downtowns have become ugly, sterile, and culturally distant. (They used to be beautiful and highly symbolic.)

A very common charge is that downtowns have become ugly and impersonal. They lack not only beauty, but also appropriate symbolism and meaning. While the centers of more traditional cities were designed as ideal urban spaces, American downtowns are more often blandly functional. In the past, downtowns were associated with beautiful architecture, pleasant parks, dazzling fountains, towering street trees, tasteful decorations, and fine-grained detail. Today, however, downtowns are made up of bland, modernistic structures with garish signs surrounded by ugly parking lots and graffiti-covered rubble. Downtowns used to be dominated by lovely churches, ornate city halls, fancy clubs, and beautiful shops while today, glass boxes and sleazy pool halls are the norm.

Well, yes and no. I have been very active over the years in the field of historic preservation and I know that a great many wonderful buildings have been torn down to make way for unremarkable glass boxes and parking lots. I also know that garish commercial signs and billboards often overwhelm our streets. I am not sure, however, that it has ever been otherwise. Photographs of American downtowns during the nineteenth and early twentieth centuries show an abundance of makeshift structures, giant signs, tangles of overhead wiring, and piles of various and sundry "products."[7] By the 1950s, primitive parking garages and aluminum siding were making their marks. Certainly some buildings from the Victorian, Art Nouveau, and Art Deco periods were very pretty, but it seems that downtowns have always been a bit of a mixed bag.

Nor does it seem that trees, flowers, fountains, and other urban amenities were all that common in most downtowns of the past. Even where downtowns were picturesque, as when cobblestone streets climbed steep hills, they may not have been that comfortable for those walking in formal shoes or riding in a wooden-wheeled vehicle. Hogs roamed the streets of New York City at mid–nineteenth century acting as semiofficial garbage collectors. American downtowns were thrown together very rapidly and they probably always had a ragged look compared to the ceremonial centers of cities in some other parts of the world.

The charge that modern and postmodern architecture is always ugly or garish is also quite debatable. While the late twentieth century is certainly different in scale and detail from earlier eras, many recent skyscrapers, convention centers, and hotels are widely admired for their beauty. The real problem is that the pieces often do not seem to fit together into a coherent whole. There are too many gaps in the urban fabric. This leads us to the second problem and myth, the disorderly appearance of the American downtown.

Problem (Myth) Two: American downtowns are ragged, ever-changing, always under construction. (They were once stable, finished, and orderly.)

Compared to many of the central cities of, say, Europe and Latin America, American downtowns are often seen as being overly functional, ragged, rapidly changing and lacking in the kind of permanent, symbolic beauty that we associate with Paris, Florence, and Amsterdam. American downtowns give the impression of being unfinished and unstable. We are unable to attach importance

to them as symbolic places that give us pride and identity. We no sooner build something than we get ready to tear it down. As the old saying goes, "They've torn down most of 'city X,' and they've put up something else." In the endless quest for profit, nothing lasts.

Once again we might ask, "so what's new?" American downtowns have always been characterized by architectural impermanence and change as well as by rapid changes in infrastructure. Old photographs show streets being widened, subways under construction, telephone poles going up, and buildings coming down. General architectural histories focusing on great American urban buildings often include statements such as "constructed in 1872 and torn down in 1906."[8] Americans pioneered the idea of the "economic life" of buildings, and that life has often been less than a half century. Perhaps the massive megastructures of the current era will be the first structures to last a century, if only because they will be too expensive to destroy.

Problem (Myth) Three: Downtowns have been privatized into fortresslike spaces that exclude much of the populace. (They were once full of more egalitarian public space.)

Many authors have suggested that today what passes for downtown public space is really private space such as shopping malls, festival market places, office building atria, theme villages, and the like.[9] These places have security cameras, private police, and a long list of rules concerning who gets in and what behaviors are acceptable. Meanwhile, the "real" public spaces of streets, parks, and plazas have been marginalized and undermaintained. Civic life has been replaced by commercial life. There is no true central place for rallies and other political events to take place. In the past, cities were more egalitarian and everyone could come downtown and take part in civic life.

While the privatization of many seemingly public spaces has undeniably occurred, I am not sure that things have changed that dramatically. It is likely that today many private spaces are nearly as egalitarian as the truly public spaces of old. It has not been that long ago that the police would arrest or at least harass anyone who did not meet the current standards of decorum. Since most people worked long hours, six days a week, few had the time to linger for long in public spaces, and loitering was viewed with suspicion. Certainly people who exhibited signs of wild incoherence for whatever reason were likely to be carted

off to the "hoosegow." Political activities were also sometimes viewed with a great deal of alarm, especially if they ran contrary to the values of the power structure; "the authorities" often ran roughshod over suspected rabble-rousers.

Women, children, and minorities were not always welcome downtown either. Unescorted women, for example, could not even eat in a decent restaurant in many downtowns, and gangs of scruffy and/or minority children were almost certain to be chased out of the heart of the central business district. In addition, African Americans and other minorities were likely to be denied service of any kind in many downtown establishments. Separate hotels and eating establishments existed "on the other side of the tracks." There were also no special facilities for the handicapped.

Many downtowns had a dearth of pleasant public space for even high-status people. The grand plazas and wide sidewalks we associate with European cities were not very common in the American downtown. The combination of street widening and overflowing commercial goods meant that sidewalks were often busy and congested. Most American downtowns were never ideal places for flaneurs or leisurely promenades.

In light of the above, it is sometimes surprising just how diverse our supposedly privatized spaces and events actually are. It is now common to see gangs of teenagers, people in wheelchairs, middle-aged couples dressed in what can only be described as an aggressively horrible manner, and people of every imaginable ethnic background all strolling through the malls, hotel lobbies, and convention centers downtown. They are being watched but they are usually let in. The courts have ruled that even political leafleting must be allowed in malls since they are the de facto public places of the modern city. Only the most disheveled of the "homeless" panhandlers are hassled, but then they have never been welcome in the CBD. Given the pervasive fear of crime in our society, it is likely that most people appreciate the levels of surveillance, especially in massive parking structures and other dark corners of the city.

Problem (Myth) Four: Downtowns have become superficial, inauthentic, and homogenized. (They once had a strong sense of place resulting from local economic traditions and architecture.)

Many criticisms of the new "postmodern" downtowns are based on the perception that they are inauthentic fantasylands geared toward conspicuous con-

sumption. As "real" economic activities have fled, they have been replaced by "historic" or themed festival marketplaces, Disney and Warner Brothers fun zones, and chain retailers and restaurants all relying on placeless architectural themes to segment the market and attract customers. Downtowns have become homogenized. You can get in a cab in any city and ask to be taken to Olde Towne, Seaport Village, or the Warehouse District, none of which have much do with being old, seaports, or warehouses. Downtowns reflect a global carnival culture that tends to divorce us from the "real," gritty city underneath.[10]

Some of these charges are obviously true. As industrial and warehouse activities have left the central city, the range in types of work found downtown has become far narrower. Stevedores and steelworkers may indeed be in short supply at the local tavern. It is also true that chain establishments have replaced many businesses that were once in local hands. On the other hand, both the trend toward homogenization and the quest for fantasyland environments have been around as long as downtowns themselves.

The invention and diffusion of downtown as a special type of place corresponded with the rise of Victorian architecture, a style that was, above all, eclectic. By the 1890s, most American downtowns featured a variety of Gothic towers, Romanesque train stations, and Greek Revival libraries. There was already competition to keep up with rival cities in the development of impressive skylines. As if three-hundred-foot towers were not enough, grand department stores, hotels, and pleasure gardens were built to cater to those who sought unprecedented fantasylands. Madison Square Garden was built to look like a Moorish castle. Downtown inauthenticity is a long-standing tradition in the American city. Perhaps this is as it should be. A city would be very limited indeed if it could only build upon the architectural and economic traditions of, say, southern Ohio or eastern Colorado.

Problem (Myth) Five: Downtowns are dangerous, criminal-filled places, especially at night. (They used to be safe, law-abiding, and full of nightlife.)

Ask a typical suburban American why he or she does not go downtown on a regular basis, especially at night, and the issue of crime is likely to come up. American central cities and crime are often linked in sensational media stories, and it is likely that many people think of big-city downtowns as unsafe. Our cities are out of control. Drug dealers, muggers, car-jackers, and bank robbers

are a dime a dozen downtown. These fears are not just the result of the popular media and folklore. Crime rates sometimes show that crime has been a big problem downtown. Part of the problem, however, is the issue of crime "rates," that is, the number of crimes in relation to the residential population. Since hundreds of thousands of people are likely to visit downtown to work or shop while only tens of thousands actually live there, crime rates are bound to be high. They tend to be high in almost every location where the daytime population is much higher than the residential population. Conversely, many bedroom communities have misleadingly low rates of crime since there may be literally no one there to mug during the day. In spite of misleading and now declining crime rates in many big-city downtowns, fear still plays an important role in critiques of downtown.

It is hard to say if crime has really increased in most downtowns over the past century. For one thing, the way we gather crime statistics has varied greatly over time and with different political administrations. We have defined crime differently too. Part of the reason for the lack of crime in the past is that fewer or at least different kinds of things were criminalized. For example, illegal immigration, prostitution, gambling, spouse abuse, and the exploitation of children were either not seen as crimes or were essentially ignored. In addition, during prohibition, a significant number of downtown visitors were, at least occasionally, involved in "drug" deals, but enforcement was usually lax. The term *tenderloin* comes from the fact that policemen in high crime areas could afford the best cuts of meat at local steak joints if they took bribes and turned a blind eye. The data are inconclusive but it is likely that crime is neither a new nor overwhelming problem in most downtowns, and it appears to be declining. Crime in the Times Square area of Midtown Manhattan fell dramatically in the late 1990s.

Riots and social chaos are nothing new either. Some of the worst riots in American history took place during the nineteenth century, as did gang warfare between ethnic groups. During the 1880s, for example, mobs burned down the courthouse in Cincinnati and would have burned more if troops had not been called in. The draft riots in New York City during the 1860s were both bloody and racist. Clearly downtowns have not always been idyllic and trouble-free in America. They have not been elsewhere in the world either.

The topic of nightlife is a different but related issue. During the early 1800s, streets were very dark and nearly devoid of life after dark. With the increasingly widespread use of gas lamps during the mid–nineteenth century, more people

went out at night, but it was not until electricity came on the scene in the 1890s that "the great white way" of dazzling signs and well-lit streets gradually became the norm. Still, it is likely that most downtown streets have always been dark and empty after the evening rush hour ended. In our collective memory, we may tend to generalize from the gaiety of a few well-lit theater districts to the downtowns as a whole.

Problem (Myth) Six: Downtowns are physically and psychologically distant— they are poorly served by public transit and parking is expensive or unavailable. (Everyone used to go downtown easily and cheaply on the bus or trolley.)

People argue that downtowns are increasingly inaccessible and remote. Most streetcar systems have long since been dismantled and only a few cities have subways or commuter rail lines. Buses, for a variety of reasons, have never been viewed as positively as rapid transit, and the situation has been made worse by cutbacks in service. They are slow, infrequent, and déclassé. (Plus, I feel that many people are sure that buses will trick them by not really going where the sign says.) Driving is not a good option either. Traffic is terrible and there is either no place to park or it is outrageously expensive. In the old days, people simply pulled up in front of their destinations; now even back streets have meters or "no parking" signs.

These are valid criticisms but perhaps exaggerated. Once cities reached a certain size, the old trolley (and bus) systems were stretched to the limit. They were very slow, often crowded, and did not always extend into all residential areas. In order to get downtown, most people had to walk quite a distance, wait at unsheltered stops, and endure a long and crowded ride. Many people never went downtown or went only on rare occasions. For much of the working class, as well as children, the elderly, housewives, etc., downtown was probably just as remote and irrelevant sixty years ago as it is today for hard-core suburbanites.

Time and distance are not the only variables, but they are important. When I was growing up, I always took the bus to downtown Columbus, Ohio, which was four miles away. The trip was usually agonizingly slow. I now realize that I could have run the distance in the same time or less, counting the time spent waiting for the bus. Today, many mid-sized downtowns are served by new light rail transit systems, but most people still prefer to drive. If anything, it may be

too easy to drive downtown. It takes only a fraction of the time the old trolleys took and parking lots and garages are everywhere, often subsidized by employers or shopping centers. We have yet to reach agreement on just how people should get downtown, but it is probably no harder today than it was in the past. We are just used to going faster. The issue of driving downtown leads to the next problem and myth, the suburbanization of downtown.

Problem (Myth) Seven: In the age of office parks and malls, downtowns have become so much like suburban developments that there is no point going there. (Downtown used to be unique; there was nothing like going downtown.)

In the age of edge cities and megamalls, downtowns have ceased to be special places. The things that we have traditionally associated with downtown, such as skyscrapers, luxury hotels, shopping plazas, fine restaurants, and cultural centers, are now nearly ubiquitous.[11] The outer belt is lined with Hiltons and corporate headquarters. At the same time, downtowns have experienced massive clearance schemes for urban renewal and highway construction resulting in the removal of anything small, old, or marginal. Downtown now consists of isolated megastructures surrounded by fields of parking, just like in the suburbs. Malls and convention centers are enclosed fortresses with little relation to the traditional street. Since downtowns and suburbs are so similar, why not just go to the closest, most convenient place? Once more, the criticism has a great deal of merit. Some downtowns have become diluted, disorganized, and placeless even as some suburban projects have become more lively and dense. But have downtowns always been that unique?

Major downtowns became big, bustling, and different only in relatively recent times. Throughout most of the nineteenth century, the kinds of buildings, spaces, and activities in most large downtowns varied only slightly from those in small towns and neighborhood and suburban centers. Typical "main streets" were lined with three- and four-story, mixed-use buildings in towns and cities of every size and description.[12]

Churches, county court houses, city halls, office blocks, and general stores all lacked size and monumentality until the later years of the nineteenth century. In addition, many small towns had bustling ports, busy railroad terminals, and a variety of craft and manufacturing activities. Beyond the confines

of New York, Chicago, and a few other huge metropolises, impressive and dominating downtown buildings were few and far between until well into the twentieth century. Most mid-sized downtowns did not have skyscrapers and huge luxury hotels until the 1920s, which was also the era of the illuminated great white ways with their theater marquees and picture window displays. It may be that downtown as a truly unique and impressive landscape really only lasted about thirty years, from the rise of the skyscraper in the 1920s to the period when the first suburban landscape elements began to creep in by the late 1950s.

Still, the suburbanization of downtown has been far from complete in most cities, and the trend has reversed in recent decades. As downtowns have retreated from the rush to widen streets, demolish buildings for parking lots, and create otherwise suburban landscapes, there is still some confusion about what to do next. Downtown landscapes have been criticized from two directions—they are too big and impersonal (stop the skyscrapers) and they are too nondescript and suburban (stop the parking lots). As a result, many cities have turned to historic preservation and the revitalization of "traditional" streets as an option. The major attractions in some big downtowns are picturesque, human-scale environments no longer found in suburbs or in "Wal-Marted" small towns.

The race to build monumental downtown structures, however, has not ended. While few new office towers are under construction these days, convention centers, sports arenas, malls, and waterfront festival centers continue to reach new heights, and new types of attractions are being invented every day. The competition is stiff, however, and even Las Vegas has a pretty good "New York skyline."

The spatial organization of most downtowns is also still very different from that of office parks and edge cities. In spite of the trend toward superblocks and inner belts, most downtowns have a tight grid of relatively small streets. This network of streets enhances pedestrian access to a variety of places. The fact that remnant buildings and functional districts still exist in most downtowns means that a wider range of rents and activities are likely to be found there than in the all-built-at-the-same-time suburban projects. In addition, there are many ways in and out of the downtown in all directions. Edge cities and other suburban developments are more likely to consist of a giant "pod" of buildings and parking lots surrounded by busy and unwalkable highways. The downtown spatial experience is hard to replicate in the suburbs.

Problem (Myth) Eight: Downtowns rely on the exploitation of low-paid "post-Fordist" labor. (They used to rely on a well-paid and well-treated working class.)

When critics discuss the growth of festival marketplaces, food courts, retail, and tourism in American downtowns, they are usually quick to point out that these service activities rely on unskilled, low-paid, nonunion labor. The food courts and hotels along the waterfront are staffed largely by teenage and immigrant employees who are willing to work for very little. The old blue-collar union jobs that once made the waterfront a good place to work have disappeared in the culture of consumption of the new downtown. Revitalized downtowns depend on the exploitation of labor.[13]

Here the notion of a past golden age is especially problematic. Downtowns, and indeed entire cities, were built upon the exploitation of labor throughout the nineteenth and early twentieth centuries. The deadly fire at the Triangle Shirtwaist Factory in New York City in 1911 demonstrated just how bad labor conditions were in most downtown factories. Pinkerton men beat up strikers through the 1930s, and nearly every working-class American was deprived of at least some goods and services by the Great Depression and the austerity of the war years. The golden age of Fordist labor only really began with the postwar prosperity of the late 1940s and 1950s, and even then most female clerical workers were seriously underpaid. Female and minority executives were seriously absent. So, just when was this period when all the downtown workers were happy and prosperous?

The idea of the traditional downtown was already under assault by the time well-paid union jobs became the norm for industrial and public employees during the 1950s. As industrial and warehouse jobs left downtowns, the remaining white-collar labor force was increasingly polarized into executive and clerical/service segments. In recent years, this polarization has perhaps gotten worse but it is nothing new. As many back-office clerical jobs have moved to the suburbs and as many high-level managerial jobs have become extremely remunerative, the typical office building has become a more affluent place. These people do not typically bring their lunches. They rely on a wide range of service people from waiters and florists to hair stylists and tailors. Some are well paid and others are not. Downtowns have always had diverse employment. If anything, however, downtown employment may be becoming more equitable, at least in terms of gender and race, as discrimination against women and minorities diminishes.

Problem (Myth) Nine: No one lives downtown; it is not a twenty-four-hour place.
(Downtown once had a large, middle-class residential population.)

This is a very common criticism of the American downtown. American cities are compared with the livelier cities of Europe, Latin America, and Asia where, normally, more people live in the central areas. As the high cost of land, coupled with purposeful slum clearance projects, has impacted downtown land use patterns, the downtown residential populations have plummeted. Downtowns empty out at five o'clock, or perhaps at six or seven if anyone stays for drinks or dinner. By eight or nine o'clock, the streets are deserted. The irony is that the most expensive land in the city is used for only eight hours a day. If downtowns are so wonderful, why doesn't anyone live there? This is a complicated question and one that deserves some discussion.

First, the twentieth-century central business district, by definition, has never really had much in the way of a residential population. The classic CBD is defined as the area given over to downtown functions such as offices and retailing and not, except in rare cases, housing. Residential areas, if they exist at all, are located on the fringes of downtown, and so the problem of downtown definition, as opposed to CBD definition, becomes very important in any attempt to measure the number of people living there.[14]

Second, since there has never been any consensus as to just what a downtown is, especially at the margins, it is difficult to measure the changing importance of a residential population. For planning purposes, many downtowns are now defined by freeway inner belts and include a square mile or more of territory. These areas often include older residential neighborhoods that were not even considered to be part of downtown only two decades ago, and these are rapidly declining in population. Can downtown be said to be losing population? On the other hand, formerly industrial waterfront districts now often feature high-density apartment and condominium projects on land that was not considered to be part of downtown in the past. Can downtown be said to be gaining population?

The question gets even trickier if we throw in social class. Proponents of the golden age remember a time when downtown was a place of proper decorum and middle-class values, yet most of the people who actually lived there were probably very poor. Downtown residences, at least near the core, were predominantly single-room occupancy hotels and flats over stores. It is doubtful

that many middle- or upper-class people ever lived in the core of most American downtowns.

It is more likely that the golden age was a time when lots of people lived close to downtown and still used its facilities on a fairly regular basis for major shopping trips, services, and recreation. For neighborhoods close to downtown, the central business district was the neighborhood center. Cities were compact until well into the twentieth century, and so it was not difficult to walk or ride the trolley from close-in communities. The suburbs had their own centers used by people living further out, but downtown still had a sort of captive audience. As inner cities have been cleared or have become poorer and older, and as automobile ownership has become more widespread, fewer people use downtown as a neighborhood center. Grocery and hardware stores were the first to close but others soon followed. By the 1970s, most of the people on the streets were part of the daytime workforce.

Things may be changing. Thanks to mixed-use projects and more flexible zoning, many downtowns are getting a significant middle- and upper-class residential population for the first time. The John Hancock Tower, which opened in Chicago in 1969, pioneered the idea of having offices and residences in the same tower, and that prototype has become increasingly popular in a variety of cities. In addition, many older downtown warehouses and office buildings have been converted to artists' lofts and other types of accommodations. Areas that are predominantly residential are still located on the margins of downtown or perhaps along a waterfront park, but these too are increasing in number. Even grocery stores are coming back.

It is unlikely that downtowns in most American cities will ever have huge residential populations. Still, as downtown populations increase and as the populations in European, Latin American, and Asian central areas thin out, American downtowns may become more like center cities elsewhere.

Problem (Myth) Ten: Downtowns are expensive toys built primarily for tourists and visitors. (They used to better meet the needs of ordinary city residents.)

All across the country there are revolts by "the neighborhoods" against what they see as excessive spending for a downtown that has little relevance to the daily lives of local communities. Highly profitable downtown activities, they charge, are run by the "downtown establishment" and by absentee chain oper-

ators largely for tourists, conventioneers, and business travelers. While immense sums are spent to get downtown ready for events such as political conventions, superbowls, NCAA playoffs, and seasonal festivals, the needs of the poor neighborhoods nearby go unnoticed. Downtown is made up of a set of unrelated megastructures such as hotel and convention facilities, sports arenas, and festival centers all aimed at attracting big spenders from beyond the city limits. It may work well as a basic industry that brings money into the city but it has no more relevance for the average person than a similarly successful steel mill.

Like the others, this critique has considerable validity. On the other hand, good cities have always been tourist destinations and much of the "life" of central cities has revolved around visitors for at least two hundred years. In Europe, the "Grand Tour" made cities such as Rome, Florence, Venice, and Paris centers of commerce and civility as early as the sixteenth century. Many of the cities of northern Spain were either founded by or blossomed with the arrival of pilgrims en route to Santiago de Compostela during the late Middle Ages. During the nineteenth century, world's fairs and expositions brought millions to the industrial cities of northwestern Europe and the United States. The cafe society depicted in paintings by the Impressionists would have been much less "impressive" without the visitors who came to Paris just to experience it.

The question becomes, how can the downtown serve both locals and tourists? The problem is complicated by the political fragmentation of most American cities. Because American central cities are often very small in area and because independent suburbs have grown up around them, many are starved for economic activity and a tax base. A large and historic downtown may be the only cash cow there is. Downtown has got to work in cities where abandoned industrial sites and burned-out ghettos make up much of the city. But sometimes it becomes a vicious circle: the poorer and more abandoned the central city becomes, the more fortresslike and irrelevant the downtown seems to city residents. At least part of the solution might lie in the realm of metropolitan political reorganization.

Still, in good downtowns, there is plenty to attract locals and visitors alike, and the mix can be exciting. Indeed, many American downtowns have become the sites for an increasing number of local events over the past decade. Jazz festivals, marathons, parades, art shows, food carnivals, seasonal celebrations, and ethnic displays have all found space on the streets and in the public spaces downtown. The highways and parking lots of edge cities cannot provide the same kinds of appropriate settings. Both locals and visitors are attracted to

these events. While visitors can be overwhelming, they can also add to the night-life of a city and provide a threshold population for a variety of restaurants and specialty shops that appeal to locals as well. Much has to do with design. If the major tourist facilities can be made to support smaller activities—the petty entrepreneurs on the side streets—then downtown can be interesting. On the other hand, if everyone stays inside fortresslike hotels and convention centers, then downtown can become irrelevant for many city residents.

Problem (Myth) Eleven: Downtowns are polluted and dirty, especially compared to the green and spacious suburbs. (Downtowns used to be clean and tidy.)

There are a few criticisms of American downtowns that are so absurd they do not merit much attention. One is that modern central cities have recently become so dirty and polluted that an escape to the suburbs is advisable. Critics contend that life is better and healthier in the clean and fresh communities far from the CBD. But one cannot reasonably argue that downtowns are more polluted today than in the recent or distant past. Old photographs of eastern cities show layers of coal smoke and soot that were sometimes so thick it was difficult to see across the street. In western and southern cities, dust was often a problem, especially when it was mixed with the droppings of the ever present, cart-pulling animals. Typhoid, cholera, and other diseases related to water supplies caused problems periodically and toxic landfills were associated with many craft industries. The past was not that clean. If anything, downtown air and water are much cleaner now than they have been for a very long time and they are getting cleaner all the time.

Today, the level of automobile and industrial pollution is largely a matter of topography and wind direction. In many western cities, for example, suburbs ringed by hills or mountains have far more serious air pollution problems than downtowns. Even so, people talk of heading up to the fresh air.

Problem (Myth) Twelve: Downtowns are crowded and congested compared to the spacious suburbs. (Downtown density was once manageable, but with the advent of cars and trucks, congestion is out of control.)

This is also a myth par excellence. Once again, old photographs show that downtowns were often so crowded that wagons and streetcars could barely

move. Whereas automobile and truck traffic may have become an increasing problem during the 1950s before freeways and inner belts began to channel truck and through traffic around the central core, traffic now moves more smoothly in most downtowns than it does along eight-lane, mall-filled suburban strips. With the possible exception of some truly high-density places like Manhattan and Boston, traffic congestion is relatively rare on downtown streets compared to past decades. The many options provided by a grid of streets makes driving less frustrating than when everyone must use the same highway, on-ramp, or feeder road. The outer belts of such cities as Atlanta and Washington, D.C., are now lined with so many office parks and megamalls that traffic grinds to a halt there at least as often as it does downtown. Congestion in America is almost entirely auto-related. The options of walking or using public transit would eliminate much of it.

The list could go on indefinitely, as nearly everyone has personal perceptions and explanations that relate to downtown decline and revitalization. In our rush to either write off the downtown as a hopeless and irrelevant venture or to proclaim a magnificent and glorious renaissance, it can be useful to ponder the possible roles of romanticized golden eras in shaping current perceptions.

Life without Downtowns: I Hope Not

Downtown is where everything comes together. It is an attic where we store our past; it is a landscape which illustrates our cultural aspirations and technological possibilities; it is a street where we meet and learn to interact with a wide variety of people; and it is a challenge that hones our skills and keeps us on our toes. Life without downtown might be comfortable in some ways, but something would be missing. To illustrate this contention and to end this section, I offer a question posed by geographer Yi-Fu Tuan: "And how can we envisage the good life (the humane life) and plan for the future unless we have some clear idea as to the sort of places that we wish to exist?"

The Evolution of the American Downtown, 1850–2000

The American downtown gradually came into being in the last decades of the nineteenth century. By the turn of the twentieth century, a recognizable and predictable kind of American city structure was becoming readily apparent everywhere.[1] For the next fifty years, downtowns became increasingly dominant, slowing down only with the arrival of the Depression and World War II. They became an accepted and much-admired aspect of the American city. Downtowns perhaps reached an apogee around 1950 with skyscraper office towers, big department stores, first-run movie theaters, and a variety of services all served by a reasonably dense network of public transit. Since then, however, the role of downtown, if not its very existence, has become problematic. In response, the second half of the twentieth century has seen attempts to reorganize and redefine both the concept and reality of the American downtown in the face of new challenges.

The Six Stages of Downtown Evolution

In order to provide some historical perspective on this thing we call downtown, it is useful to outline some of the major stages or eras most downtowns

have experienced. The six stages of downtown evolution presented below provide a framework for understanding the most important processes in the creation of the American downtown.[2] They are: (1) inception, (2) exclusion, (3) segregation, (4) expansion, (5) replication, (6) redevelopment.

During the inception stage, downtown first begins to emerge as a distinctive place, even though there are few, if any, special types of buildings associated with it. The homes and coffeehouses of the elite often begin the process, serving as meeting places for important business transactions. Only the oldest American cities experienced this stage since, by the mid–nineteenth century, cities were normally laid out with designated "business blocks." During the exclusion stage, the economic and land uses that are most obviously not appropriate for a booming downtown are forced away by a combination of high rents, social pressure, and architectural change. In other words, when special-purpose office buildings and stores began to define the new downtown, many uses could no longer afford to compete for space there.

Segregation occurs when downtown functions are sorted out on the basis of both rent and type of required interaction. Typically, the larger the downtown, the more segregation exists.[3] In the largest cities, there may be special areas for legal services, banks, government activities, retailing, and so on. All of this requires expansion, and downtowns have usually needed to grow in area or vertically, or both. In most cases, areal expansion is contiguous, while in some cases, growth must "leapfrog" unsuitable locations. Similarly, vertical growth may be unbridled or building ordinances regulating the height and bulk of structures may control it. The replication stage may be reached when expansion becomes difficult or when suburbanization becomes so prevalent that there is a need for one or more replicas of downtown elsewhere in the metropolitan area. Finally, successful downtowns are rebuilt continuously so as to compete for the highest-level activities and most prestigious buildings. This is known as the redevelopment stage. To this list of accepted stages, we might add a seventh: the stage of reinvention or redefinition of downtown.

In recent decades, most downtowns have become something far different from the central business districts of old. In spite of decades of homogenizing trends and consensus goals with regard to such things as architectural styles, uniform building codes, transportation, and housing, downtowns have retained individual personalities and spatial arrangements. In this book, I am after both the predictable and the unique. I seek both a general model of the American downtown at the millennium and an appreciation of each down-

town's sense of place. I divide this chapter into two sections, the invention and evolution of the downtown during the period from 1850 to 1950, and the renewal and redefinition of downtown over the period from 1950 to 2000.

The Early American Downtown: European Antecedents

The story of the emerging downtown can be organized around six topics of discussion: European antecedents, the development of special-purpose architecture, changing modes of transportation, cultural values, downtown morphology, and the political-economic system. The first of these, European antecedents, will have to be especially brief, as the concept of a central business district took several centuries to develop in a large number of places.

The specialized business district probably originated in Europe during the later decades of the seventeenth century, but it developed very slowly there. The weight of past traditions was too heavy and the truly specialized downtown did not come into its own in most European cities until well into the 1800s. In America, the downtown evolved more quickly since most of the components were in place as cities were first organized. Except in the very oldest American cities, such as Boston and Charleston, S.C., railroads, department stores, factories, and office buildings all played a role in the early years of growth and development. In Europe, however, these components took hold only very gradually.

Many of the characteristics that we think of when we imagine a good downtown originated in Europe. Among these are monumental buildings, grand boulevards, specialized business and shopping districts, and vibrant commercial waterfronts. The composite image, however, was a long time in the making. During the mid-1600s, the City of London, for example, had perhaps 100,000 people living in its one-square-mile territory. It, like all the rest of the cities of the world, was a city of mixed uses.[4] Special-purpose architecture had not yet been developed to any real degree and people usually lived and worked in the same buildings. The scale of cities was very different then, and rarely did an urban area encompass more than about two square miles, an area that has been referred to as the walking city. No place was very far away from anything else, and so streets were narrow and pedestrian-oriented.

Since cities were very small in area and most craft endeavors could easily be integrated into the traditional built environment, there was little need for specialized economic zones. In the medieval cities of Europe, the segregation of

land uses and economic activities was often intermeshed with patterns of residential segregation. Since people lived and worked in the same buildings or neighborhoods, weavers, tinsmiths, and dyers usually had their own craft districts. Because the center of the city was the best place to do business, it was also the best place to live. In Islamic cities, for example, the most valuable goods were sold in the very heart of the bazaar, protected by a social and architectural labyrinth from intruders. Activities thought to be noxious or unpleasant, such as tanning, were located in peripheral settings and/or downstream. Although the very centers of large cities were dominated by sacred structures such as cathedrals or mosques, market squares and covered bazaars brought commercial activities close to the core as well.

The first hints that the spatial organization of cities was changing appeared in the late seventeenth century. The one-square-mile City of London was depopulated by the plague of 1665 and the fire of 1666 and never really regained its status as a place to live. As London approached a quarter of a million in population, medieval patterns, already disrupted by the disasters, began to break down. Elite residential squares were pioneered beyond the decaying city walls, and those who could afford it commuted to businesses in the City from the new West End. A specialized downtown was slowly being invented.

Other changes in scale were being explored in the Baroque cities of Europe such as Paris and Rome. Grand tree-lined boulevards were introduced during the seventeenth and eighteenth centuries, along with monumental palaces, bridges, fountains, and other urban decorations. The closed and vertical world of the medieval city gave way to the horizontal patterns of the Renaissance. Long vistas and a sense of perspective became important. As the city opened up, things became further apart, and a good location close to the center of activity became more than just symbolic in importance. Those with access to carriages could travel longer distances, although most did not. By the late eighteenth century, social and economic segregation was increasing in the larger cities of Western Europe. This idea would diffuse rapidly during the century to follow. In London, for example, financial activities concentrated in the core city while government offices, upscale retailing, and housing located in the West End well away from the warehouses and working-class East End.

A key point here is that downtowns or central business districts did not really exist as recently as 1800. We are dealing with a relatively new urban phenomenon. A second point is that as we have attempted to revive downtowns in recent decades, we have often turned to the past for ideas. We have tried creat-

ing monumental public spaces, elite residential districts, enclosed shopping bazaars, grand boulevards, busy waterfronts, urban fortresses, and even modern versions of the city wall, all dating from the preindustrial, pre-downtown city. The past may be important as a guide to the future.

For the most part, the early American cities of the Eastern Seaboard borrowed architectural traditions and patterns of spatial organization from the cities of northwestern Europe. There were, however, some significant differences right from the beginning. The most important difference pertains to initial planning and the use of a grid. While some of the earliest American cities, such as Boston, evolved with an irregular street system reminiscent of medieval times, others were planned from the beginning around some variation of a grid. Savannah and Philadelphia, for example, had grids from the time of initial settlement, and New York City adopted the grid in 1811. In addition, few American cities ever really had walls for very long, nor were they built around a permanent, sacred core.

American cities tended to be street-oriented rather than place-oriented. Rather than clinging to a plaza major or cathedral, prestigious economic activities sought a location on the "main street." American cities were more linear than their antecedents. People and things were always on the move, up Broadway in New York or Charles Street in Baltimore. When they became fully developed, American downtowns tended to be linear as well. So the American city was a hybrid, partly European and partly invented on site.

The Development of Special-Purpose Architecture

The main component of the traditional central business district is business, and so the specialized downtown as an urban district had to wait for the evolution of both business activities and the function-specific buildings that housed them. This was a very gradual process. As early as the sixteenth century, a few European cities had developed exchange halls as centers for conducting high-level business transactions. By the late 1600s, banks and warehouses began to cluster around these halls and form a district that was known to be a lively center of business activity. Still, these areas were quite mixed, with barrels of rum or fish, inns, small factories, and a variety of sheds and shanties all on the same street. Much business still took place on the wharves or in coffeehouses. Until well into the nineteenth century, most buildings remained multipurpose, with little if any difference in outward appearance between a house,

a warehouse, a banking house, or a schoolhouse. Gradually, however, architecture became much more function-specific, and "business blocks" were built on the major streets that were being defined as "downtown."

By the middle of the nineteenth century, the centers of American cities were characterized by having a wide variety of special-function buildings. These were only prototypes, however, since the early office buildings, department stores, exchange halls, tenements, and warehouses found in and around the emerging downtown tended to be relatively uniform in appearance.[5] Still, the stage was set for the explosion in architectural specialization that occurred after the Civil War. By the 1890s, there was no mistaking the individuality of the office skyscrapers, luxury hotels, and shopping arcades that characterized the new American downtown. The central areas of European cities remained more uniform in appearance and mixed in land use, although there, too, changes were afoot.

It was during the early decades of the twentieth century that American downtown architecture was perfected.[6] A combination of technological breakthroughs such as geared elevators and steel frame construction meant that skyscrapers could be built to unbelievable heights. Electric lights, central heating, and escalators meant that huge department stores could be constructed so that vast interior spaces could be comfortably used by hordes of shoppers. A variety of modern appliances made life in high-rise apartments not only possible but also luxurious, and theaters, movie palaces, skating rinks, bowling alleys, and train stations all became increasingly specialized and recognizable. As buildings became larger and more specialized, land uses became more segregated. In the mixed-use city, everything was everywhere. By the early 1900s, particular activities were located according to their ability to pay for a location on expensive, centrally located land. The concept of the rent gradient had arrived in full force. As buildings became more specialized, activities became more segregated and people had to travel longer distances to work and shop. By the 1920s, these patterns were being institutionalized through zoning regulations.

Changing Modes of Transportation

As long as cities were compact and nothing was more than a mile or so away from anything else, most everyone walked to their destinations. By the 1920s, however, many downtowns were as large as the entire city had been a century earlier, and people needed some new devices to travel to and from them. For-

tunately, there were an increasing number of options. During the 1850s, the horsecar emerged as the earliest form of reasonably rapid transit. As the name suggests, they consisted of horse-drawn vehicles pulled along iron tracks. The advantage of horsecars over earlier carriages and omnibuses was that they were not only faster and more comfortable, but they were also larger and cheaper. Because tracks diminished the considerable friction that had existed with cobblestones and mud, fewer horses could haul more people longer distances for a lower fare. Thus, residential suburbs could be built further out. But the advent of fixed tracks had another very important impact on city structure, the development of predictable high-volume locations along the tracks, especially where the tracks converged.[7]

The place where the tracks came together became known as the peak land value intersection, or PLVI. This was the best place to locate businesses that relied upon high visibility and large-threshold populations. This was the core of the emerging downtown. Although the city stretched ever further into the countryside, people still converged on the center from all directions. During the 1890s, the horsecar lines were gradually electrified and the streetcar or trolley continued the pattern of the star-shaped city with a strong core. Streetcar riders peaked during the 1920s, but this remained the primary mode of transportation into American downtowns until the late 1940s, when trolley buses, motorized buses, and automobiles began to take over the streets of the city. For nearly one hundred years, fixed tracks and a strong city center evolved together.

In a few of the largest cities, the demand for transit into the downtown was so great that the streets could not accommodate it. Elevated transit lines and subways evolved to facilitate the development of multilayered transportation. Without these layers, the massive downtowns we see in cities such as New York and Chicago would not have been possible. European and Asian cities also developed streetcars and subways, but there older spatial patterns often preceded the transportation innovations and it was not easy to change them. Ironically, the multinodal city so much associated with America really is more typical of the older cities elsewhere in the world. In London, for example, business concentrated in the City, government at Westminster, and entertainment in the West End. A dense network of transportation was created, but no consensus focal point exists.

The city of tracks and nodes began to break down with motorized transportation, especially with the widespread ownership of the automobile. By the late 1950s, most people were looking for places to go where they could park

easily and cheaply. But city structure does not change overnight. For several decades, suburban commercial developments remained inferior to those downtown, and by the time they reached comparable levels, total infatuation with the automobile and car-designed environments had begun to wane. Drive-in movies and restaurants had passed their primes.

Perhaps the most controversial dimension of American urban transportation has been the expressway or freeway. Limited access highways were not new in 1956 when federal highway funding became available for 90 percent of the cost of a national system of freeways. Partly justified by the needs of national defense, freeways linking the major cities (and military bases) of the country were to be built using a federal tax on gasoline. For cash-starved cities, it was free money, and highways could also be located so as to clear slums. Over the past few decades, superhighways have been built throughout the world, but nowhere have they had the impact that they have had on American cities. In Europe, autobahns and autostrada connect various parts of nations, but nowhere do they slice through the hearts of cities to the degree that they do in the United States. As Robert Moses said as he designed new highways for New York City, when you have a dense urban area, you have to hack through it with a meat ax.[8] Freeways per se are not necessarily bad for cities, but the "meat ax" approach usually was.

Still, the complications resulting from a constant stream of big trucks through the heart of downtown was a major component of the negative image of the "congested" central city. On the other hand, the re-routing of all through traffic to inner and outer belts contributed to the sense of emptiness in some downtowns, especially those with wide streets. In the early days of freeway construction, limited access highways were often part of a pro-downtown strategy. One of the arguments for early freeway hub-and-spoke arrangements is that they would allow for easier and faster commuting in and out of downtown. Such highways have served to do just that in many parts of the world. However, in the United States highway planners went a step further.

While highways in cities are not unique to the United States, two important spatial arrangements are: the inner belt and the outer belt. Both of these features have had a major impact on downtowns. Inner belts are freeway loops that go all or nearly all the way around downtown so that through traffic need not bother with city streets. The problem is that they have tended to isolate downtown from the residential neighborhoods beyond. In addition, most central business districts are much smaller in area than the space enclosed by a typ-

ical inner belt. Consequently, a sort of no man's land exists between the downtown core and the highway that is neither fish nor fowl. Downtown as defined by an inner belt usually includes vast but underutilized zones of remnant houses, parking lots, and old warehouses. The "normal" city of houses, churches, and grocery stores usually begins well beyond the inner belt. Downtowns have become psychologically distant from the rest of the city, especially where old railroad yards or port facilities reinforce the highway barrier. Elevated freeways and ramps are particularly problematic from a design standpoint, but in all cases, access to the downtown is limited to occasional crossover or under streets.

Significantly, several cities, such as San Francisco, Portland, and Boston, have recently eliminated some of their downtown freeways or are putting them in a tunnel underground. Also significant is the fact that many of the more successful downtowns such as Denver, New Orleans, and Boston do not have inner belts. Surprisingly, the impact of these inner belts has not been systematically studied from the standpoint of urban morphology and spatial connectivity.

Outer belts have also impacted the downtown, albeit from a distance. By connecting the spokes of the freeway system with a circumference highway, new points of high accessibility have been created out where land is relatively cheap and available in large parcels. When such nodes are embedded in an affluent residential sector, the result is often the creation of a massive "edge city" or midtown comparable to the old downtown.[9] Cities that have tried to limit suburban growth, such as Portland, Oregon, have an easier time of it if they do not have a complete outer belt.

The bloom may be off the extreme infatuation with highways and auto-dependent strips. For one thing, land has escalated in price since the 1950s and the cost of new urban highways has reached ludicrous levels. There have been cultural changes as well. By the 1980s, drive-in restaurants and movies on the auto-created commercial strips had largely given way to more pedestrian-oriented experiences in malls and historic districts. Things may yet come full circle. Much depends on cultural values and the resulting policies.

Cultural Values and Landscape Tastes

The topic of cultural values is necessarily broad and amorphous, but to a very real degree societies create the kinds of landscapes they prefer. For example, Americans have been accused of being anti-urban with regard to residen-

tial preferences, and so we have garden suburbs with rustic, neo-half-timbered houses surrounded by white picket fences. But we have other tastes as well. At least until very recently, we have valued big, new, clean, flexible, formless, and technologically advanced landscape features and disdained the old, small, stable, antiquated, weathered, and orderly landscapes so often found in other parts of the world. Cultural preferences go a long way toward explaining our somewhat chaotic, ever-changing, skyscraper-filled downtowns. We have been fickle citizens of our cities as we have constantly required new landscape features to keep us amused. As long as downtowns provided the best and the newest, we revered them, but when suburban megamalls offered more glitz, we were quick to change allegiance. It is not always simply a matter of convenience, as people will drive long distances to get to a major attraction.

American cities usually lack the kinds of permanent cultural symbols that provide a sense of stability to central cities elsewhere in the world. We have rarely had the castles, cathedrals, central plazas, palaces, city gates, monuments, or even the grand boulevards that have anchored central activities in other cities. We have tended to avoid becoming attached to stable, historic landscapes capable of providing an enduring urban sense of place. In the American downtown, the most prestigious location is often where the newest, tallest, most impressive building happens to be at the time. While general location and street address can still be important, being in the Chrysler Building or the Transamerica Tower often says more. For over a hundred years, corporations have competed to create the new center of the CBD simply by building the biggest, most famous address in town.

The Political-Economic System

A variety of government policies aided and abetted the striking contrast between the highly developed core and underdeveloped frame of the typical American downtown. During the early decades of the twentieth century, a variety of zoning regulations and building codes were put in place that sometimes served to worsen conditions in the marginal areas of downtowns. Many downtowns, for example were overzoned in the sense that allowable building heights and volumes greatly exceeded the existing situation. The temptation was to let the older, smaller buildings decay and eventually replace them with much larger, more profitable structures. In contrast, European cities usually limit the size of new buildings so that they more closely match the existing

urban fabric; it is worth maintaining older structures when nothing larger is allowed.

In addition, it became fashionable in American cities to have single-use zoning so that commercial, industrial, and residential activities could be strictly separated. This eliminated flexibility and made it hard to convert buildings that were obsolete for one function to a more appropriate one. Hardships were inevitable. When marginal commercial buildings could no longer have apartments upstairs, owners were deprived of needed rent and often had to close the street-level store as well. Good intentions often led to vacant buildings, not to mention the loss of needed housing.

Building codes also caused a variety of problems for older buildings in the zone of transition. Fire, seismic, and many other building codes and regulations were a reaction to the frequent disasters that befell the teeming cities of the nineteenth century. The idea was to encourage the demolition of old and dangerous buildings and to replace them with modern, fireproof structures. After all, this had happened in many European cities as early as the 1600s when all buildings were required to be of stone or brick construction. In American cities, however, two problems quickly became evident. First, as skyscrapers became fashionable, the demand for new space was taken up by a relatively small number of large buildings rather than being distributed throughout the central city. New York did not emulate Paris. Second, the streetcar and later the automobile allowed for the construction of endless commercial strips and competing business nodes. The result in each case was diminishing demand for marginal downtown buildings.

Financial institutions exacerbated the situation. Buildings that did not meet the current code could not get financing. Owners wanting to upgrade or even maintain buildings had trouble getting loans, and sales required huge down payments, as mortgages were difficult if not impossible to obtain. Laws required that if a certain percentage of a building's value was spent on repairs or upgrading, then the building had to be brought all the way up to code. This was usually unreasonable since codes were often arbitrary with regard to, for example, banister heights or window coverage and because rents could not be raised very much in marginal buildings. Insurance was difficult to obtain for these older buildings as well. Meanwhile, everything was easy in the suburbs. For a tiny down payment and a low-interest mortgage, a larger space could be had for less. Only gradually did some cities begin to realize that they had made slums almost inevitable. Anti-redlining legislation and flexible code and zoning enforcement appeared only during the late 1970s.

Downtown Morphology, 1900–1950

By the 1920s, the combination of skyscraper office towers, luxury hotels, massive department stores, and immense theaters meant that American downtowns were becoming more compact even as they were growing in size and importance. In most middle-sized downtowns, the activities of the CBD were confined to a handful of major buildings on no more than a dozen or so blocks, while older buildings on the periphery experienced disinvestment. These older buildings became part of what has been called the zone in transition or, in extreme cases, skid row. Typically, they were used for marginal activities or torn down for surface parking as owners waited for the day when the CBD would expand in their direction and the land could be sold for the construction of a major building. In the meantime (often for decades), typical downtowns consisted of gleaming towers surrounded by acres of deterioration: a core surrounded by a frame. With the coming of the Depression and World War II, this slow but dynamic process froze into a seemingly permanent downtown characteristic.

In reality, most downtowns had two distinctive types of zones of transition because the peak land value intersection (PLVI) or core of the CBD tended to move in one general direction over time. Usually, new and prestigious buildings were constructed uphill in the direction of wealthy residential neighborhoods and away from the industrial waterfront or railroad yards. The area into which downtown grew is referred to as the zone of assimilation, while the area that was abandoned is the zone of discard. The zone of assimilation typically features old houses, stores, and churches while the zone of discard is the home of old industrial structures, warehouses, and outdated hotels and office buildings. Both types of districts experience decay and disinvestment for the same reason—the expectation of change. Thus we use the term *transition zone*. Speculation in land was coupled with disinvestment in buildings as property owners awaited a better day—a day that was usually quite slow in coming.

Creating the "Modern" American Downtown, 1950–1975

Downtown was still by far the major center of economic and social activity in the early 1950s, but the combination of implosion and decay had already taken a toll. When people speak of going to a typical middle-sized downtown during the 1950s, they are usually referring to a very small section of what we

now call the downtown area. The big department stores and movie palaces were normally concentrated on one or two streets, as was the sense of a busy, vibrant center city. First-class office buildings occupied a few more nearby streets, but the CBD was still quite compact. When one went downtown, it was usually to Euclid Avenue in Cleveland, Wisconsin Avenue in Milwaukee, or Peachtree Street in Atlanta. It was here, and on nearby side streets, that downtown life flourished. Not far away, old buildings and old businesses hung on, as much due to the lack of options as anything else. When the (suburban) option appeared later in the decade, the margins of most downtowns thinned out and the number of vacant buildings and surface parking lots increased dramatically. This last golden age of the traditional downtown was short indeed.

With the passage of the National Housing Act in 1949 (amended in 1954), the popularity of massive urban renewal schemes financed in large part by Washington increased measurably. The Interstate Highway Act of 1956 added more federal dollars for the purpose of clearing slums, and the urge to tear down became nearly irresistible. While many cities did have real slums ripe for demolition, the main focus of urban renewal and highway construction during the 1950s was in and around downtown. Beginning in Pittsburgh and diffusing rapidly, large sections of American downtowns were scraped clean. By the early 1960s, many of these "slums" had been replaced with gleaming modernist office towers rather than with housing—symbols of the rebirth of the tired old American city. The designs were typically modernist in the extreme, with glass and aluminum boxes set back from the street or on pedestrian podiums atop parking garages. The number of street-level doors plummeted, often going from twenty or more per block to one, usually a bank lobby. Freestanding, aesthetically unrelated buildings replaced the more coherent street walls of the past. In other cases, cleared areas remained vast grassy fields for years.

Some cities experimented with pedestrian malls even as they were clearing away much of the housing and commercial activity that would support them. In general, downtowns had become less walkable by the 1960s. It has become very fashionable in recent years to criticize both urban renewal and modernist architecture, and I, for one, am very happy to do so. However, one very important caveat may be in order. By the mid–twentieth century, American downtowns were in need of the shock of the new. Many downtowns were polluted, worn-out, and periodically under water. Coal smoke was so thick in downtown Pittsburgh that streetlights were sometimes kept on all day. About half of downtown Cincinnati was under water during the flood of 1937, and the docks

and warehouses around Baltimore's inner harbor were crumbling. In the newer cities of the South and West, the situation was less bleak but problematic just the same. Most cities were simply underbuilt, with tiny cottages and tourist motels next to modest downtown buildings. In nearly every city, it was felt that it was time to have a new downtown fit for the modern age. Something had to be done. Many of the mistakes of the early modernist era can be viewed a bit more charitably as wobbly first steps on the road to the redefinition of downtown. Still, many horrendous mistakes were made.

Another relatively charitable view of the massive clearance schemes of the 1950s and early 1960s evolves from the realization that most leaders at that time believed that lots of new buildings were needed in a big hurry. The economy was booming and people were demanding new, air-conditioned space. In those days old stuff was not charming, it was just old, especially if it was an office building or a hotel. There was still too much of it around for it to have any scarcity value. Suburban locations were still not a viable option for prime office space and major retailing in the mid-1950s (although the first malls were beginning to open). If everything new was going to go downtown, space had to be made available, so buildings were cleared away to make room for the expected new construction.

The trouble was, everything did not go downtown. While there was a boom in downtown office space construction during the 1960s and early 1970s, nearly everything else had either already left the central city or would do so soon. Indeed, the excitement of seeing the new office supertowers going up (still many of the tallest in the United States), may have blinded civic leaders to the problematic nature of increasingly one-dimensional city centers.

Taking Stock of the Postwar Boom

In some ways, the worst was over by 1975. The twenty-year period of massive highway and urban renewal schemes was drawing to a close, along with the disinvestment in targeted areas that such projects caused. Downtowns were on the road to a level of permanence and stability that they had not seen for a generation. The new skyscrapers brought a solid and relatively affluent workforce that gradually led to the establishment of a variety of food courts and small retailers. Most cities had new gleaming skylines and plenty of new highways to bring people to see them. There were problems, however, at street level. The combination of the loss of smaller, peripheral buildings due to obsolescence

and clearance for parking, along with the construction of massive, one-per-block skyscrapers in the core of the downtown, meant that there were fewer street-level doors than ever before. Would-be pedestrians encountered seas of surface parking while making their ways to glass boxes surrounded by sterile plazas.

Most modernist design principles either ignored the street or purposefully avoided contact with it. Superblocks, plazas, podiums, and sky bridges were all the rage. At all costs, people were to be kept away from the dangerous and smelly street. These principles worked very well. Soon, pedestrians looked out of place. Blank walls punctuated only by occasional gaps for underground garages replaced street walls of individual buildings with stores. Waiting for a bus became a bleak and lonely experience.

Most downtowns had imploded into a relatively small number of blocks by the mid-1970s. The old fine-grained and variegated urban fabric had been replaced by a handful of office towers that, even in mid-sized downtowns, sometimes had as much as one million square feet of space. Disinvestment continued in older, class B and class C office buildings as owners still expected eventual demolition. Department stores, when they continued to exist at all, offered more downscale products aimed at close-by central city markets, as if suburban shoppers would never again be attracted downtown. More than ever, many downtowns took the form of a modernist and somewhat sterile core surrounded by an underutilized frame.

The underutilized frame often seemed a bit scary. The increasing use of graffiti to decorate unloved and uncared-for marginal properties meant that much of the peripheral downtown was defaced. Derelict and abandoned buildings attracted marginal uses at a time when the media began to concentrate on the issue of rising crime and drug usage. Movies such as *Escape from New York*, *Blade Runner*, and *The Out-of-Towners* played up the dangers of urban life. Indeed, the word *urban* became a euphemism for a variety of dangerous and wicked settings. The image of the American downtown was at perhaps an all-time low during the 1970s.

Creating the Postmodern Downtown, 1975–2000

Obviously it is hard to pinpoint an exact date when things began to change, especially since significant events happened at different times in different cities. It is also difficult to know just when new kinds of projects were first conceptu-

alized and planned, since bringing them to fruition often took years. Still, 1975 seems like a pretty good date for recognizing new trends in urban design for American downtowns. For one thing, the recession of 1975 brought a symbolic end to 1960s-style projects and provided a breather from rampant development. Gradually, planners and developers became more familiar with different ways of thinking about cities.

In 1971, the *Urban Design Plan for San Francisco* was published and was widely heralded for its inclusion of new ways of viewing planning and design.[10] The plan suggested, among other things, that urban design be more grassroots and it encouraged inputs from the potential users. Borrowing Kevin Lynch's ideas from *The Image of the City*, the plan focused more on how ordinary streets should look and feel than on clearing land for massive projects.[11] As a result, micro-level urban details were given greater importance. Fine-grained street walls were in and monumental superblocks were out. Books such as *Defensible Space: Crime Prevention through Urban Design* suggested that buildings did not need to be built to look like impenetrable fortresses in order to be safe.[12] In addition, the successes of Quincy Market, Pioneer Square, Ghirardelli Square, and a variety of other preservation projects received acclaim when the Bicentennial was celebrated in 1976. During the mid-1970s, anti-redlining legislation became widely accepted, and cash for mortgages and maintenance began to flow to older, marginal buildings for the first time in decades. Skid Row gradually became Olde Towne as brewpubs in old warehouses were "invented."

As summarized (albeit tongue-in-cheek) by Tom Wolfe in *From Bauhaus to Our House,* total infatuation with the "less is more" look of modern architecture faded as the 1970s drew to a close.[13] After nearly twenty years of impersonal glass boxes with sterile plazas, it was time for a change. New types of towers began to appear, such as the pyramidal Transamerica Tower (1972) in San Francisco and the "Chippendale" AT&T Building (1980) in New York City. Planners began to realize that their good intentions as manifested in zoning regulations were part of the problem. Throughout the 1960s, for example, New York City gave zoning bonuses, or additional allowable space, for the provision of open plazas at the base of buildings. The result, after a decade or so, was often one empty plaza after another and the elimination of any possibility for a lively street wall of shops and cafes. Gradually, the rules were changed to encourage buildings with upper-level setbacks and street-level retailing. Instead of hiding from the street, buildings were brought forward to it and decorated with lively facades.

In part due to changing architectural fashion and in part due to changing zoning and height and bulk regulations, the simple glass box began to give way to skyscrapers with tops reminiscent of the Art Deco towers of the 1920s and 1930s. Terra-cotta buildings with stepped facades and rounded or pointed tops gave urban skylines more colorful personalities. The rules had changed; by the early 1980s, many cities were getting more lively buildings from top to bottom. At street level, they were more pedestrian friendly, and "upstairs" they were more colorful and even whimsical since they "referenced" or even mimicked the architecture of the past. This new (old) architecture has been dubbed "postmodern" since, while it returns to many of the time-tested aesthetic guidelines of the past, it is largely a reaction to the excesses of modernism. Some have critiqued the new style as inauthentic, however, because the buildings are in most respects quintessentially modern but clothed in a facade of tradition.

Some critics of the late-twentieth-century American city argue that this facadism is the key characteristic of today's urban landscape and that everything is weirdly inauthentic. While there is some merit in this argument, I feel that it is greatly overdone. Throughout urban history, colorful, decorative, and even fanciful architecture has given life and visual interest to the streets of cities. It would be hard to argue that nineteenth-century Greek Revival plantation houses in Mississippi or Victorian castle-inspired train stations in Indiana were any more "authentic." The admirers of the postmodern hold that "less is more" was a bore and that downtowns are finally getting dressed up again.

The changing styles ushered in by the turn toward postmodernism coincided with the tremendous economic and building boom of the 1980s. As a result, many of the more vibrant downtowns were transformed very quickly with lots of new buildings and districts. Even mid-sized downtowns experienced millions of square feet of new development, as well as significant amounts of recycled space in preservation-adaptive reuse projects. The sheer volume of all the new stuff gave many downtowns a psychological boost. People, and capital, were apparently interested in downtowns once again.

Removal of the Poor and the Changing Image of Downtown

While there is little doubt that planning and architecture generally became more sensitive and colorful after 1975, some other aspects of the downtown renaissance were more problematic. For one thing, after years of urban renewal and highway schemes aimed in part at "poor removal," in many cities the poor,

or more precisely their residences, had finally been eliminated from downtown. The immediate result, of course, was an increase in "homeless" populations as old single-room-occupancy (SRO) hotels were destroyed and small above- and behind-shop flats were disallowed by zoning regulations. The poor paid the price for the middle-class fear of crime, disgust with graffiti and drug dealing, and the general feeling that "urban" areas were out of control. Downtowns needed to be "cleansed." Gradually, laws against aggressive panhandling, public urination, and vagrancy were enacted and/or enforced in many downtowns.

In addition, many nearby "inner city" neighborhoods thinned out as a result of a combination of slum clearance, upward mobility and improved housing choices for minorities, and gentrification. By the 1990s, many downtowns were no longer perceived to be shopping districts for the poor and hangouts for the down-and-out. But the trend has had uneven results. Some downtowns, such as Los Angeles, have retained huge "skid row" districts while others, such as Boston, have experienced widespread gentrification. Elsewhere there are significant variations. Seattle and Cincinnati still have large, low-income neighborhoods downtown while Indianapolis and Columbus have relatively few. Portland has a high-profile skid row but it is rapidly diminishing in size. In Cleveland, the east side "inner city" has thinned out dramatically in recent years. In spite of the very real concerns for "homeless" populations, however, most mid-sized downtowns are now less likely to be associated with "urban slum" conditions than they were only a few decades ago. Like it or not, the path has been cleared for the arrival of middle- and upper-class residences.

Recapturing Waterfronts: The Importance of Physical Site

Next to the new skylines and historic districts, the most obvious and universal trend in American downtowns has been the conversion of long-ignored and underutilized physical features into major amenities. While the largest and most famous American cities have long had major physical amenities—such as New York City's Central Park, Chicago's Lakefront, and the Boston Common—few mid-sized American downtowns have made similar use of their physical sites until quite recently.[14] Indeed, the riverfronts of such cities as Cleveland, Milwaukee, Minneapolis, New Orleans, and Portland were major toxic disamenities until well into the twentieth century. In other cities, such as Indianapolis, Phoenix, and Denver, rivers were all but invisible except when

they were at flood stage. Lakefronts were problematic as well. Downtown Cleveland shied away from pollution-filled Lake Erie and Milwaukee ignored Lake Michigan as it expanded westward. As cities have gradually removed outmoded industrial structures and port facilities, immense amounts of waterfront land have become available for recreation.

In recent years, many of these "water features" have become major amenities. Parks, boat rides, biking and jogging trails, and waterfront restaurants and cafes have been developed in order to give downtowns the types of attractions that cannot easily be replicated in the suburbs. The image of downtown as a center of outdoor recreation represents a major shift in the definition and role of the area we still refer to as the "central business district." While most cities still retain a balance of functions, it is likely that more people go to downtown San Antonio to stroll along its Riverwalk than go there to work in offices.

The Search for Big Buildings and Space-Extensive Land Uses

After decades of implosion, many downtowns have recently increased in size due to a variety of massive, space-extensive developments. Chief among these are ballparks and sports arenas, convention centers, educational and medical complexes, and housing developments. While in many cases there is a perceived real need for the new structures, in others it may be primarily a matter of finding things to soak up the vast territories made available over the years through slum clearance and deindustrialization. Sports arenas and associated parking, for example, not only take up a lot of space but can spin off other activities such as sports bars and souvenir shops. Similarly, convention centers, taking up several acres of space, can spin off hotel, entertainment, and restaurant facilities. Big-footprint buildings can give scarcity value to the areas nearby.

In some downtowns, huge areas have been set aside (zoned) for high-density or even suburban-style residential complexes. Finally, medical and educational institutions have enlarged and expanded their campuses in and around many downtowns. New universities have been located downtown as well, such as the Indiana-Purdue complex in Indianapolis and the multischool Auraria campus in downtown Denver. After years of clearance that left vast, empty fields, many downtowns are gradually beginning to fill up.

The trick is to insert these new space-extensive activities in such a way that they contribute to downtown spatial structure rather than diluting it. Big-footprint buildings, for example, can separate and isolate activities and deaden

downtown space if they are located in the wrong place. Massive convention and sports complexes inevitably have long, blank exterior walls which contribute nothing to the streetscape. They cannot be located right in the middle of things. On the other hand, if they are too far away from the core of downtown, there are few spin-offs or complementary uses. The same goes for housing. If residential units are included on the upper levels of mixed-use projects, they can easily be inserted into the heart of downtown, but large, homogeneously residential complexes can have a deadening effect on downtown street life. In part, this is due to the emphasis on security in most market-rate housing. Very few residential projects have doors on the street. Many have a variety of walls and gates meant to deter city-resident interaction. It remains to be seen whether such features will remain the norm as downtown residences become more common and accepted.

Back to the Future: Reviving Mass Transit

Big cities have long been dependent on a wide variety of mass transit. The spatial structures of New York City, Boston, and Chicago would be inconceivable without the variety of subways, elevated lines, and light rail found in those cities. For much of the past half-century, however, most middle-sized cities have relied on automobiles and, to a much lesser extent, motorized buses to get people in and out of their downtown areas. During the halcyon days of freeway construction and urban renewal, there seemed to be plenty of room for the automobile as the new off-ramps led to acres of surface parking. In recent years, however, things have begun to change.

While only a handful of cities, such as Washington, D.C., San Francisco, and Atlanta, have built new heavy rail and/or subway systems during the past two decades, light rail has become popular in a variety of mid-sized urban areas. Portland, Sacramento, San Diego, Baltimore, Denver, Dallas, and Cleveland all have new and/or greatly expanded light rail systems connecting the downtown with suburban residential areas and other major nodes of activity. Other cities, such as Seattle, Minneapolis, Charlotte, and San Antonio, have revitalized bus systems, sometimes complete with new downtown central stations. In successful downtowns, parking is becoming more expensive and less convenient. While it is presently too soon to tell, it is likely that light rail and buses will play major roles in shaping the spatial structure of downtowns in the not-too-distant future.[15]

What's Next? Project Planners vs. Facilitators

Those who seek to revitalize American downtowns often have been polarized into two very different camps with two very different approaches. On the one hand, there are those who favor monumental projects that will attract a lot of attention and give potential residents and investors confidence in the future of downtown. On the other hand, there are those who argue that big, slow-moving projects may cause more harm than good. This latter group favors facilitating as many small, incremental improvements as possible. Downtown needs what Jane Jacobs once called "gradual money" rather than cataclysmic change.

Perhaps a good compromise between these two extreme positions is that downtowns should progress in stages or cycles, alternating between big projects and gradual upgrading. The trick is to keep a balance so that different districts experience change at different times and at different rates. For example, it could be that after years of urban renewal and skyscraper construction, we are now entering an era in which smaller improvements such as street trees and comfortable bus stops merit attention. Of course, different cities and different districts within cities are at varying points along the continuum. At the "turn" of the millennium, for example, New York City is focusing on such things as controlling pornography, improving streetscapes, rehabilitating old buildings for new uses, and even courteous behavior. On the other hand, it is nice to be able to do these improvements in settings that resulted from big projects— such as the ice rink at Rockefeller Center.

The jury is still out on just what constitutes an ideal scale and rate of change. While many European cities have long concentrated on gradual improvements with little significant architectural change, few American downtowns have had a comparable stock of beloved older buildings. Indianapolis is not Florence. It may be, with the massive developments of recent decades, that we have finally achieved a building stock big enough and expensive enough to be worth preserving. On the other hand, it could also be that Americans simply like change and that downtowns will forever require a degree of novelty.

The Downtown Stage

Physical Site, Street Morphology, and Civic Space

Urban scholars have long written about the role of physical site in the initial location and subsequent growth of cities. Site refers to the physical geography at and immediately surrounding a settlement. Site characteristics include such things as soil, vegetation, climate, topography, harbors, bays, rivers, and other water-related features. Traditionally, the most attractive sites were those with good harbors, lots of trees for building and fuel, and plentiful supplies of water for irrigation, rich soil, and sometimes, an easily defended position. As American urban areas have sprawled over tens and even hundreds of square miles in recent decades, it is only the downtown that still occupies the original physical site. These sites often give personality to the entire city even though they are no longer the settings for port activities or natural resource extraction. Urban personality and sense of place result from a combination of attributes and flaws. The most memorable settings are those that provide stunning visual images but also a certain topographic awkwardness. Flat, easily developed areas, such as those often found in new suburbs, are usually less problematic but more easily forgotten.

New York City is an example of a city with a good site. Manhattan Island

provided a reasonably defensible position in the early years and was also blessed with plentiful supplies of water and timber and good, well-drained soil. Most importantly, the harbor is large, deep, and protected. By the mid–nineteenth century, New York City had become the dominant port in North America, handling a majority of U.S. foreign trade. Success was also due, of course, to situation. New York is midway between the northern and southern cities of the East Coast and had access to the interior via the Hudson and Mohawk river systems even before the construction of canals, railroads, and highways. New York, like Boston, was also relatively close to the trading centers of Europe. But even today, site contributes greatly to the image of New York City.

Obviously, physical site characteristics can change over time. Harbors can be filled in, hills sluiced down, and water piped in. Many medieval European cities started to decline as a result of river or harbor silting. In addition, sites that are thought to be perfect in one era may be viewed as obsolete during another. Defensible hilltop sites were valuable during medieval times but were much less useful in the age of nation-states, cannon, and Renaissance planning ideals. Many had become backwaters or museum towns by the mid–nineteenth century.

In recent years, much of the literature dealing with urban growth has ignored site and focused instead on the changing "situation" or relative location of cities with regard to trade areas and transportation networks.[1] Discussions of Central Place Theory, the growth of the Sunbelt, globalization, and regional economics are more common than those on physical site characteristics such as waterfalls and swamps. Part of the reason for this is scale. Major metropolitan areas now cover hundreds of square miles and have widely varying site characteristics. Los Angeles, for example, includes mountains, deserts, beaches, vast plains, and a variety of microclimates within its sprawling metropolitan borders. When cities were smaller and more compact, the importance of physical site was easier to discuss.

A second reason for the emphasis on studies of relative location is that situational characteristics are easier to analyze using quantitative measures. Statistical studies of trade flows, communication networks, migration streams, and the like all relate to a city's situation. It is harder to measure the impact of, say, a sunny climate or a topographical setting that allows for good views. In addition, the assumption that site can be greatly modified using modern technologies has become widely accepted. By the twentieth century, site could be "conquered" by sluicing down hills, filling in harbors, and building massive highway

bridges. Physical settings, it was thought, could be obliterated or at least built over in order to achieve economic and social priorities. In part as a reaction to some extreme contentions by environmental determinists, scholars began to assume that people could do what they wanted wherever they wanted to do it. Given enough money and willpower, the exigencies of place could always be overcome. As a result, the role of physical site in the study of urban change has been largely ignored in recent decades. By the 1960s, only the tourist literature made reference to the hills of San Francisco or the bay islands of Miami.

Physical Site and the History of Urban Form

While even historical studies have tended to emphasize the goals and perceptions of people over the constraints and opportunities afforded by different types of physical sites, in the past, such constraints and opportunities did play more obvious roles. In Classical Greece, for example, city builders sought sites for an acropolis or high city which could be used both as a defensive retreat and as a highly visible setting for sacred structures. Cities dominated by hilltop castles were common for thousands of years and in a variety of cultures from Salzburg to Himeji. Historians and geographers often classified cities on the basis of their origin around particular kinds of physical sites. There were, for example, portage sites where major rivers could be crossed (Paris), head of navigation sites where rivers flowing to the sea became navigable (London), confluence sites where major rivers came together (Pittsburgh), as well as cities that grew up around harbors, narrow valleys, peninsulas, coal deposits, and the like.

Although cultural values were of utmost importance in the design, spatial organization, and architecture of cities, the physical site was always something to contend with. Cities such as Athens, Edinburgh, Stockholm, Istanbul, Dubrovnik, Boston, San Francisco, New Orleans, and Mexico City provide examples of places where cultural priorities evolved gingerly in challenging physical settings. Each of these cities retains a unique sense of place very much related to its site characteristics. While not all cities have been similarly challenged, site has almost always been a factor in the spatial organization of cities.

During the height of the age of industrialization during the late nineteenth and early twentieth centuries, the physical sites of cities were often either bulldozed into submission or covered over with so many factories, railroad lines, and warehouses that they were barely visible, especially in heavy smoke. Site characteristics gained attention mainly at times of crisis such as during floods,

earthquakes, or severe air pollution. The physical site was more often something to be conquered or used up than something to be appreciated. Harbors were dredged, trees were cut down, and coal was dug up. While a good physical site could bring advantages to a city, it often had to be pounded into submission in order to be properly utilized.

There were, of course, some exceptions to this generalization. With the rise of leisure time and tourism around 1900, people began to seek beautiful and serene settings for relaxation. Most of these were in the countryside—antidotes to the big city. Many, however, gradually grew into sizable cities themselves. Miami Beach, for example, grew primarily because of the perceived potential attractiveness of its site. It became a major destination for tourists by the 1930s. Los Angeles also offered a variety of attractive sites, especially for the emerging movie industry. Mountains, deserts, beaches, and agricultural valleys were all only a few miles away. Filmmakers moved from New York in part to take advantage of these varied physical sites.

Since the site of the initial founding of a city is most often where the downtown is today, it is there that site characteristics are the most visible. American downtowns tend to be either along the shores of a river, harbor, or lake and/or nestled in valleys or narrow coastal plains. Physical site and downtown identity often go hand in hand.

Physical Site and City Sense of Place

In recent decades, physical site characteristics have begun to play a major role in downtown revitalization. Most of the postwar plans for revitalizing downtowns were relatively placeless in the sense that the goal was to create "modern" landscapes and efficient functional arrangements. Cities tended to skirt around the issues of sense of place and the character of the physical landscape. Victor Gruen's plan for downtown Fort Worth (never carried out) epitomized this approach.[2] It had as little to do with the city's location on the cowboy frontier of Texas as possible. Other cities, however, attempted to combine physical features, geographic location, and cultural heritage in order to create more placeful central areas. Utilizing Depression-era public works, San Antonio, for example, enhanced its downtown image largely by rehabilitating its small and sleepy river. As the city prospered in the postwar decades, it continued to focus on its "Spanish" Riverwalk and the tourist appeal of the Alamo. It emphasized the fact that it was a southwest Texas town.

Perhaps the first city to make at least a halting attempt to capitalize on its physical site as an integral part of urban design planning and city image-(re)making during the immediate postwar era was Pittsburgh. Downtown Pittsburgh occupies a stunning physical site, nestled between high bluffs where the Allegheny and Monongahela Rivers come together to form the Ohio River. By the late 1940s, however, its site was more of a negative characteristic than a positive one. The rivers were polluted and subject to periodic flooding. Downtown was often under water. Air inversions and the steep hills combined to keep the smoke and grime of nearby industries trapped in the valleys. Sometimes it was difficult just to see across the street, and lights were kept on at noon. Pittsburgh was not an attractive place.

The killer floods of the 1930s and the killer fogs of the 1940s convinced the leadership of Pittsburgh that something had to be done. Since the city is home to a number of large corporations, there was considerable financial clout behind the plan to essentially reinvent the image of the city. After going full steam ahead during the war years, Pittsburgh had entered a long and continuing period of economic and demographic decline in the decades that followed. As the steel and related industries closed, many saw a revitalized downtown as the only hope for a sagging economy. Pittsburgh would have to change from a blue-collar to white-collar city in order to stem the tide. To a very real degree, Pittsburgh pioneered the idea of site-specific downtown revitalization through the linking of urban design and physical setting.[3] While it was not the first city to utilize this idea (the Chicago Lakefront, for example, was earlier), it was the first big city to attempt to remake its image by publicizing the beauty of its physical site. In the years since, scores of cities from Baltimore to San Diego have followed suit.

Pittsburgh invented the phrase "the Golden Triangle" when it cleared the various and sundry railroad and highway bridges, storage sheds, factories, and slums that occupied the Point—the confluence site where the three rivers meet. By the early 1960s, the Point itself became a green park with a monumental fountain, while gleaming glass and steel skyscrapers were built just behind it. A decade later, Three Rivers Stadium was built just across the Allegheny, and a new image for the American downtown gradually came into focus—water (complete with boat rides) in the foreground with a green and inviting waterfront flanked by office towers and sports emporia. Downtown Pittsburgh was no longer seen as a "pit" but rather as a place to be emulated.

The revitalization of downtown Pittsburgh also coincided with the emer-

Pittsburgh's Golden Triangle and downtown skyline

gence of the "view of the skyline" as an important dimension of real estate pro-
motion. Although the idea of a city view was not new—after all New York City
had long been famous for its skyline—in most cities, a view of downtown was
not considered to be much of an attribute. When Pittsburgh's downtown was
ringed with a tangle of industrial landscapes usually enveloped in smoke, there
were few who would pay very much to look at it. Even without the heavy industry and smoke, few mid-sized downtowns were visually stunning before the
office boom of the 1960s. Boxy brick buildings of six to ten stories may have
had pleasant facades up close, but they were rarely appealing from a distance.
Rooftops were generally flat and contained massive water tanks and other functional accouterments. As downtown Pittsburgh was reinvented and rebuilt,
luxury apartment buildings appeared across the river in Mount Washington
and Duquesne Heights because the new city offered something to see. The idea
quickly diffused and a skyline view became a symbol of urbanity in the American city from Boston to Seattle.

   While Pittsburgh remade its image by focusing attention on its physical site,
by the 1960s, its success was only partial. Expressways and interchanges marred
connections between the downtown and the rivers. Aside from the Point, the

riverfront was more likely to have parking lots, highways, and railroads than pedestrian access. The city really began to focus on riverfront revitalization only in the 1980s, but its first efforts were seminal nevertheless.

Waterfront Reclamation and the New Downtown

As epitomized by the 1954 film *On the Waterfront*, shorelines have traditionally been known as part of the work-a-day world, the underbelly of urban society. The first signs of change began in Europe more than a century ago. Cities such as London, Paris, Florence, and Rome had built embankments and walkways along at least some sections of their centrally located rivers by the mid–nineteenth century. While warehouses and fish markets remained, rivers became aesthetic attractions as well as transportation routes. Indeed, some have argued that the Renaissance began in Florence when the middle section of the newly built "Ponte Vecchio" (1345) was left open in order to provide a view of the Arno River. In the area of waterfront design, North American cities were far behind those of Europe until quite recently.

The first aesthetic rumblings along North American waterfronts occurred with the City Beautiful Movement of the early 1900s and the WPA and PWA projects that followed during the Depression. Still, the efforts were few and far between. With the exception of such projects as the Chicago Lakefront, the Columbus, Ohio, riverfront, and the new government complex on San Diego Bay, most urban waterfronts were largely ignored by the civic improvement schemes of the pre–World War II years. Indeed, many waterfronts became uglier as the twentieth century wore on. In cities such as Boston, Cleveland, Seattle, and Portland, massive highways joined existing rail lines and industrial zones along the waterfronts, thus further separating the downtown core from a potential, but unrecognized, amenity. In most cities, the downtown simply migrated away from the "industrial" waterfront during the early decades of the twentieth century. From Milwaukee to San Francisco, the center of downtown activity moved ever further from the "zone of discard" along the waterfront as the decades passed. This has changed in recent years.

The emerging consensus is that "water features" of some kind are a nearly essential component of downtown revitalization. They give the downtown a sense of focus and often a sense of history as well as providing attractive settings for evening strolls or major festivals. This puts some cities at a disadvantage. A few downtowns, such as Atlanta and Charlotte, have no water-

fronts to work with. Others, such as Denver and Indianapolis, are struggling to establish connections with rivers that traditionally have been physically and psychologically distant from the downtown core. Wherever it is possible, however, downtowns are seeking to reestablish some kind of linkage to water. Indianapolis, for example, has removed railroad tracks and reinvented an old canal in order to link its downtown with the newly expanded White River State Park. Cleveland now has a waterfront park and marina complementing its Rock and Roll Hall of Fame. After avoiding any unnecessary contact with Lake Erie and the occasionally flammable Cuyahoga River during the decades of hyperpollution, the city is making its lake and riverfront important parts of its persona.

Their locations along and historical association with important rivers, lakes, harbors, and bays give most downtowns an advantage when it comes to competitive place-making. Selling the downtown as a place to invest increasingly means selling it as a memorable place. The list of waterfront revitalization projects is almost endless; it includes such diverse cities as Boston, Baltimore, Savannah, Miami, New Orleans, St. Louis, Cincinnati, Portland, Seattle, Sacramento, San Diego, Spokane, and Knoxville. It remains to be seen whether cities

Cleveland's Lake Erie waterfront

with more challenging river connections, such as Phoenix and Dallas, will be able to follow suit.

Hybrid Waterfronts: The Quest for Fun and Function

While there has been little controversy associated with the goal of cleaning up American lakes and rivers, there has been some concern over the loss of "authentic" activities on the waterfront. It may be, of course, that it is difficult to have it both ways. In order to make Lake Erie live again, some of the heavy industry had to go. Still, if all of the traditional activities associated with water are removed, there is a danger that the downtown will offer little more than a suburban edge city with commercial attractions built around an artificial lake. Cities, of course, do not always have a choice in the matter. Some waterfront activities simply cannot be preserved. Small, shallow harbors can no longer compete for big ships, and local fishing grounds may decline in productivity. Even the navy may decide to consolidate in other locations. But if downtowns are to maximize the value of their physical sites, a viable mix of old activities and new attractions would seem to make sense. Cleveland and San Diego are cases in point.

Downtown Cleveland grew up around the place where the Cuyahoga River flows into Lake Erie. Now known as "the Flats," this once heavily industrial area has been somewhat cleaned up and is the location for a number of cafes and entertainment venues. Upriver, however, a number of industries remain and must be serviced by ship. As bar patrons sit on decks overlooking a sanitized but still industrial landscape, freighters several city blocks long carefully make their way up the meandering river. One false move and the Improv Comedy Club would be history. In spite of its being updated, the Flats does have a strong sense of place and function that cannot be replicated in a mall or edge city.

For the past eighty years, San Diego has been a navy town. Over ten square miles of the central city in and around downtown belong to the military. The "traditional" waterfront activities that in most cities would be industrial and port-related have never existed to any extent. Indeed, in spite of its excellent harbor, San Diego does not even rank among the top one hundred ports in the United States, and, if anything, it is becoming less important. Even the tuna fishing industry with its related canneries is on the wane. But aircraft carriers, destroyers, submarines, and a variety of other military vessels abound. To some degree, the navy has become a tourist attraction with "open" ships available for visits

Cafes along the Cuyahoga River in Cleveland's "Flats" district

and shows put on by Navy Seals and skydivers. This tends to soften the military landscape and make it part of the local sense of place, although some problems remain. The noise and pollution from some (especially air) craft can be bothersome, and many worry about the potential problems associated with nuclear-powered submarines and carriers. On the other hand, the San Diego waterfront seems destined to remain a "place" that cannot be replicated in the suburbs.

Downtowns everywhere face a similar challenge—how to make the physical setting as attractive and appealing as possible without losing identity, uniqueness, and sense of place. Physical settings, especially waterfronts, usually provide both problems and opportunities in the revitalization of downtowns.

Topography and Identity: Hills, Valleys, Cliffs, and Islands

Water, although perhaps of utmost importance, is not the only physical site characteristic that has shaped downtowns. Rugged topography has also played an important role in local sense of place. While San Francisco is the most famous example of an American city known for its hills, topography is also an important element in the downtowns of, to name a few, Cincinnati, Portland,

Seattle, San Diego, Pittsburgh, and Providence. Hills can provide a sense of physical and psychological separation between "incompatible" land uses in and around downtown. In San Francisco, for example, elite apartment buildings and hotels have long existed only a few blocks but uphill from skid row. In Portland, some of the most expensive residential areas in the metropolitan area are located on hills overlooking downtown. Similarly, the Mt. Adams section of Cincinnati has been a pleasant retreat from the downtown and river just below for over a century. Complex topography can thus be a factor in the maintenance of social and demographic diversity downtown.

Topography can also "suggest" where certain activities should be located. Without wishing to seem an environmental determinist, I must say that it is far more common for railroads and heavy industry to be located in valleys between residential hilltops than vice versa. The expanding cores of central business districts have also often been constrained, or at least "topographically challenged" by difficult terrain. The point here is that not all downtowns are alike and that sometimes topography continues to be a major factor in decisions affecting major land uses. Even after decades of urban renewal and plans to make it otherwise, the Hill District next to downtown Pittsburgh remains primarily residential.

Physical Site and the Size and Shape of Downtowns

Downtowns vary greatly in area and compactness. There are, of course, many historical and functional reasons for these variations, including era of initial settlement, types of transportation systems, and purposeful design and regulation. But site plays a role as well. The compact downtowns of Boston, Pittsburgh, and Seattle occupy very different physical settings than the more sprawling central areas of Houston and Phoenix. In attempting to delimit, describe, and generalize about downtown structure, it is important to consider how physical settings have helped to shape the central city. Most downtowns occupy relatively challenging sites compared to the edge cities of the suburbs. Suburban nodes were often purposefully located on the flattest, least complicated sites available in order to minimize the costs of development. Consequently, these suburban centers are often much larger in area than traditional downtowns, and most have been designed for easy expansion over time. On the other hand, it is hard to find downtowns that are similarly uncomplicated. Even when the land is nearly flat, winding rivers, flood plains, and small but awk-

ward topographical variations have sometimes led to interesting spatial patterns downtown.

Downtowns are thus full of "edges" and "breaks," that is, abrupt and significant changes in land use and visual character.[4] While such breaks have often been problematic in that they have made downtown expansion difficult, they can also provide real and potential visual excitement. Thus waterfronts and hillside view lots have attracted increasing attention in downtown revitalization. Often, natural breaks have been reinforced by later decisions such as those resulting in the location, size, and spatial organization of major parks, industrial districts, and residential areas. The street pattern of the downtown also reflects the original physical setting. Often, the original grid ran parallel to a river or lakefront while subsequent additions were oriented to cardinal directions, thus forming "mismatched" grids. While there can be a number of reasons for breaks in a grid, when such breaks are reinforced by physical features, the impact on downtown shape can be significant.

Climate and City Life: Reviving a Forbidden Topic

After all of the energy required to refute the sometimes absurd statements of the extreme environmental determinists who wrote about the impact of climate on human societies, urban scholars have tended to ignore the topic in recent decades. With the renewed interest in "city life" and the downtown as a place for festivals and spectacle, however, some discussion seems appropriate. As the role of cities has changed and as more downtown functions depend on voluntary rather than required activities, the vagaries of climate can take on new importance. When and where outside activities are an absolutely necessary part of urban life, they can and do take place in all types of climates, from Stockholm to Calcutta. Indeed, in sections of American downtowns where activities are based more on serving basic needs than on the leisure time of the affluent, the sidewalks are likely to be full of life. From Broad Street in downtown Newark to Broadway in downtown Los Angeles, the streets are normally lively and colorful in all kinds of weather.

There is little doubt that climate does not *determine* human behavior. A problem, however, arises when we are discussing discretionary activities—activities that are entirely voluntary and are based more on pleasure than survival. Here too, people do persevere in all types of climates when there are few options. The Javanese still dance when it is hot and Finns go shopping all win-

ter. The problem is that many of our images of vibrant city life are based on the leisure activities that evolved in Europe from the late eighteenth to the early twentieth century. When we rhapsodize about the pleasures inherent in a revitalized downtown, we often focus on promenades along busy streets lined with sidewalk cafes, flower markets, beautiful fountains, and street performers. We condemn empty streets as lifeless and anti-urban. When there is no one "outside" enjoying these amenities, the downtown has failed. While the emergence of city life was far more related to changing economic, social, political, and architectural relations than to physical environment, the fact remains that most of the major European cities are located in relatively benign climates. Walking, chatting, and even sipping coffee can take place outside for much of the year in Rome, Lisbon, and Paris. Many of the architectural traditions that evolved in Europe during this period, such as arcaded streets and public squares, were aimed at facilitating "public" life outside. We often borrow many of these architectural characteristics when we create urban design plans that depict how the "good downtown" should look and function.

In North America, climatic conditions are likely to be more extreme than those in Europe. Whereas cross-country skiers may enjoy the weather, few urbane flaneurs want to be outside for very long in Minneapolis or Cleveland in January. Similarly, sidewalk cafes are not a big attraction in Phoenix when the temperature passes 110 degrees Fahrenheit. Given the choice, most people would rather be inside. Obviously "culture" can overcome climatic considerations, but it is not always easy or convenient. People frolic in the snow-filled streets of Quebec during winter festivals and the people of Madrid have adjusted to the summer heat by eating dinner at eleven p.m. For busy Americans used to air-conditioning and central heat, malls and gallerias make for stiff competition when we are deciding where to spend our leisure time. When evaluating American downtown design, we must be careful not to be too Eurocentric. Skyways in Minneapolis, for example, are a problematic but not entirely bad idea.

It may be that climate is occasionally a factor in measuring the success of European-inspired revitalization efforts. For example, glass-covered bus stops combined with required street-level retailing have worked very well in Portland, Oregon, where the climate is mild though sometimes soggy. They may not be as successful in Houston or Phoenix. Still, projects like the Riverwalk in San Antonio demonstrate that, given enough design planning, historical association, and major attractions, people will still walk around outside even when the weather is oppressive. It could be that it has only been the combination of

The Riverwalk in San Antonio

oppressive weather and oppressive design that has killed many American downtowns.

The Rating System

The following rating system is an attempt to evaluate each of my sixteen case-study downtowns on the variable of physical site and use thereof. The possible scores range from 1 (D) to 10 (A). I have decided to be reasonably charitable in my scoring by adjusting it to what might actually be possible in a typical American downtown at this point in time as opposed to a utopian image. Since the downtowns I studied all show some signs of success, none scores below a C−. Similar rating systems will follow the discussions of each of the ten variables.

Variable 1. Physical Site and Sense of Place

Portland: A (10)

There are excellent connections between the downtown core and the Willamette River (a National Heritage River) all along Tom McCall Waterfront Park. The park was created during the early 1970s when a controversial river-

The Willamette riverfront in Portland

side highway was removed thus stimulating a major change in nearby land uses. The river provides aesthetic and functional diversity and a sense of history and open space where active "port" activities and grain elevators exist alongside marinas and waterfront cafes. There is also excellent use of trees and other vegetation in the many downtown park blocks. Hillside parks and green residential areas enclose the downtown on two sides, making for a "cozy" and picturesque site. Stunning, although occasional, views of Mt. Hood help provide a sense of place as well.

Seattle: A (10)

Seattle has an exceptionally long and varied waterfront including a large area still used for port facilities. Closer to the downtown core, there are excellent and varied uses of waterfront buildings and old piers. Lots of recreational boats as well as fishing craft add life to the scene. Waterfront parks and view sites (good views of islands and mountains to the west) help to attract amenity-oriented projects like condos and restaurants to the center city. Topography adds interest to downtown since there is a steep slope upward from the waterfront to the east. Sometimes this acts as a barrier, however, since it can serve to separate the downtown core from the waterfront, making the journey seem difficult. Hills

The Seattle waterfront and marina

also add pizzazz to the skyline, since the tallest skyscrapers are on much higher ground than the waterfront.

San Diego: A (10)

Historically the core of downtown was disconnected from the bay, especially since dredging and landfill projects during the early 1900s created large under-utilized tracts of muddy land. Now, however, connections are much improved, with projects such as Seaport Village and Marina Park providing promenades and recreational destinations. However, the now-expanding convention center tends to wall off the water from much of downtown, as does Pacific Highway, parking lots, and rail lines further south. Navy ships and tuna boats still serve to make the waterfront interesting and functional. Views of Coronado and Point Loma add interest, as do boat excursions and ferries. Modest hills within downtown add variety, which along with nearby Balboa Park, serves to reinforce the northern edge of downtown and discourage sprawl.

Pittsburgh: A (10)

The original Golden Triangle was defined by the river confluence at Pittsburgh's distinctive "Point." The downtown core is surrounded by hills (and

Downtown San Diego with the bay and Point Loma beyond

even cliffs) and water, making it an extremely distinctive and compact place. The park and fountain at the confluence along with "in progress" parks along and across the three rivers have improved access to the "new and cleaner" physical site, though Point State Park is a little off-center for most downtown visitors. Boat excursions ferry people to sports venues across the river, and there are recreational excursions as well. Views over the city from Duquesne Heights help to give the city a memorable iconography. Nevertheless, the scene is marred somewhat by the history of flood dangers, since the problem, although greatly ameliorated by federal projects, has not been completely solved. As deindustrialization occurs, downtown activities are expanding to occupy both shores of the once polluted Allegheny and Monongahela Rivers.

Baltimore: A (10)

Like most port cities, Baltimore turned away from its cluttered harbor and the core of the downtown moved up Charles Street throughout the late nineteenth and early twentieth centuries. As heavy industry and port facilities moved seaward during the 1960s and 1970s, the harbor was gradually cleared and redeveloped as part of an expanded downtown. While some might argue

Baltimore's Inner Harbor and downtown skyline

that the harbor is really beyond the classic "CBD," it is now part of the downtown recreational complex. Downtown is being wrapped around the harbor, and boat rides (rentals and shuttles), historic "waterfront" districts, and bayside promenades give the city much of its character and identity. The downside is that much of the traditional downtown away from the harbor has suffered relative neglect in that the harbor site is used as the chief icon of the city.

San Antonio: A− (9)

Probably the best example of creating an excellent physical feature from a marginal attribute, the Riverwalk provides the main focus for downtown activities, and it is being expanded continuously as a focus for various projects and attractions. The San Antonio River is little more than a stream, and its depth, width, and course have been so greatly altered over the years that it is difficult to say whether it qualifies as a "natural" or man-made feature. It is even drained and cleaned each year. Still, the river and its trees and landscaped banks provide the city with much of its personality. Rather than being a stunning and monumental site like those of Seattle and Pittsburgh, it provides a shady and cozy refuge from the Texas plains.

Cleveland: B+ (8)

Cleveland grew up at the spot where the Cuyahoga River flows into Lake Erie, but by the late nineteenth century, both were so industrial and polluted that downtown moved eastward along Euclid Avenue and turned its back on the water. In the 1960s, the river was so laden with pollutants that it actually caught fire and burned. The Lake Erie waterfront also came to be occupied by railroad tracks and a municipal airport. Today, the Cuyahoga River (a National Heritage River) has been revitalized as the "Flats," an industrial area evolving into an entertainment district. The character is still intact—complete with Lake Erie steamers winding by the cafes. The lakefront has been reconfigured as a recreational "harbor" with the Great Lakes Science Center and the Rock and Roll Hall of Fame. Boat excursions help to attract visitors to the water. However, the river and its associated incised valley acts as a barrier to downtown expansion and physical access to the west and south, and the lakefront has yet to be completely connected with the downtown core.

St. Louis: B+ (8)

St. Louis is a classic riverfront city that once had a large downtown along the banks of the Mississippi. The core of downtown moved away from the river during the late nineteenth century, leaving behind a historic but decrepit skid row. By the 1960s, most of this had been cleared to make way for the Gateway Arch and a major highway. Today, there is a potentially grand riverfront park and an adjacent small historic district, Laclede's Landing. There are also casinos and boat excursions to attract people to the water and give identity to the city. Still, the entire area remains somewhat underutilized and disconnected from the downtown core—a passive retreat as much as a focal point. The downtown is still somewhat physically and psychologically separate from the original site, but connections are gradually improving and the riverfront is used for a variety of recreational activities.

Columbus: B (7)

Columbus is located at the confluence of the Olentangy and Scioto Rivers. There is nothing spectacular about the site, but there is a lot of riverfront property since the Scioto winds through the downtown core, which is situated mainly on the east bank. Since flood-control projects have lessened the dangers of a location on the west side, new projects are being built across the river, notably the new Center of Science and Industry (COSI). Riverfront parks are used heavily for special

The Scioto River and downtown Columbus

events, with the city skyline as a backdrop. Plans call for riverside trails and other "waterfront" features. Though small, the river is part of the city's personality. Civic Center Drive, which follows the river, is often used for festivals and special events.

Providence: B (7)

Three rivers converge in and around downtown Providence, but over the years they have been lined with industries and railroad tracks and even covered over with highways. In recent years, they have been "uncovered" and embellished with parks and walkways and even filled with festive torches. A new mall now bridges the river with Venetian-like gondola rides as an attraction. In addition to water, the downtown core is surrounded by hills that add visual character by facilitating views of the state capitol and nearby historic residential districts. However, the physical site, although cozy, makes for a very constricted CBD with little room for needed expansion.

Minneapolis: B (7)

Located near the headwaters of the Mississippi River, Minneapolis has long made waterpower for mills a part of its persona. Nevertheless, though rapids

The new Waterplace Park and Riverwalk in downtown Providence

and locks provide a tremendous potential attraction, the city has historically turned away from the river. Today, there are parks and heritage walkways, but the riverfront remains an off-center retreat, although the "north shore" has experienced some major development as deindustrialization has occurred. Lakes also loom large in the city's personality, and Loring Park and Lake to the south of the downtown core acts as a major focus for residential and cultural activities.

Denver: B− (6)

When people think of Denver, its stunning mountain backdrop usually comes to mind, but few physical attributes actually define the downtown core. The South Platte River is disconnected from the city core by a vast and as yet largely undeveloped field on the other side of the railroad tracks, although plans are under way to correct this by eventually creating a new downtown expansion zone in the area. Cherry Creek has been revitalized and is lined with pathways, but the downtown core has turned its collective back (especially that of the convention center) on it. Ironically, other than scenic distant views, the capital of the Rockies has a relatively unfocused, although pleasant, physical site.

Rapids on the upper Mississippi in downtown Minneapolis

Indianapolis: B− (6)

Historically, the downtown core has turned its back on the White River with railroad lines and slum housing between the city and the pleasant but small and unnavigable waterway. In recent years, the connections have been improved through a combination of land clearance for such things as a zoo, an art museum, and a university (Indiana University–Purdue University at Indianapolis), along with the reconstruction of an artificial canal linking the river and the city. Still, there is the sense that the river is "over there" rather than a core downtown attraction. The canal, however, like the San Antonio Riverwalk (which it emulates) is helping to provide a focus for residential development at the edge of the CBD. Eventually, downtown may gain a better sense of connection to and integration with its riverfront.

Charlotte: C (4)

The downtown core has no real discernible physical foci since there is no river or other water body or any real topographic definition. The downtown core follows a low ridge, but this only serves to accentuate the distance between the office core and the marginal uses beyond and so provides little benefit, al-

though the fact that the core is not completely flat does add some visual interest. On the other hand, there are no physical obstacles or disruptions to hinder downtown interaction.

Atlanta: C (4)

Atlanta is perhaps America's largest downtown with no signature physical features or foci. The small canyon or "gulch" that provided a low setting for initial rail development contributes to the walling off and compartmentalization of the core's activities, although much of it has now been covered over for deck parking. As with Charlotte, the office district occupies a ridge which makes the CBD excessively linear and disconnected from nearby downhill support districts. The ridge, however, does serve to enhance views of the skyline from a distance.

Phoenix: C− (3)

There are no physical attributes in downtown Phoenix other than the usually dry but occasionally threatening (flooding) Salt River corridor. The riverbed is most often seen as a disamenity and a bleak barrier between the city core and the communities to the south. There are pleasant views of distant mountains, but the downtown core lacks both a consensus focus and any sense of pleasant boundaries. As a result, core activities tend to drift off in at least three directions.

Summary: Physical Site

Portland 10
Seattle 10
San Diego 10
Pittsburgh 10
Baltimore 10
San Antonio 9
Cleveland 8
St. Louis 8
Columbus 7
Providence 7
Minneapolis 7
Denver 6
Indianapolis 6

Charlotte 4

Atlanta 4

Phoenix 3

Street Morphology and Spatial Organization

The street network defines the basic spatial organization of cities. While most studies of downtowns concentrate on architecture and building types, the spaces between buildings matter as well. These are the spaces that provide the settings for much human activity as well as for movement. While physical site characteristics, such as topography, can influence the size and arrangement of streets, there are many other factors to consider. For example, tiny passageways and cul-de-sacs characterize the medinas of the older cities of North Africa and the Middle East. These are pedestrian streets designed to limit interaction and increase privacy. There is a world of difference between these types of settings and the eight-lane highways and freeway ramps that are common in many American suburban areas. Most American downtowns feature a grid of average-sized streets and so lie somewhere in between these two extreme examples of street morphology.

Variations on the Grid

With the exception of some of the very oldest American cities, such as Boston, most American downtowns are built on some variation of the grid plan. Even so, there can be tremendous variations. Some cities, such as Columbus, have just one downtown grid while others, such as Atlanta, have several grids meeting at awkward angles. Some cities, like Salt Lake City, have wide streets and huge blocks, while others, such as Portland, have narrow streets and small blocks. A few American cities, such as Washington, D.C., and Indianapolis, have radial street patterns and traffic circles superimposed on a grid. Some cities have seen their street systems greatly modified through the creation of superblocks and expressways, while others have constructed special transit lanes for buses or light rail. Many downtowns have alternating one-way streets, while in others, it is still possible to drive around the block.

The number and variety of basic street arrangements is almost endless. Some downtowns have alleys running parallel to the streets, thus providing areas that can be used for deliveries, parking, cozy pedestrian lanes, or trash receptacles. Others have no alleys and must give over a portion of the building

facades to garage and service entrances. Many downtowns have widened streets, usually at the expense of sidewalk space, while others have done just the opposite. Some downtowns have encouraged on-street parking or even head-in parking as a buffer between pedestrians and traffic, while others have as many lanes of fast-moving traffic as possible. There are city streets that have benefited from extensive tree-planting schemes and give over a significant part of the public right-of-way to greenery, while others are dominated by commercial advertising.

The infrastructure of telephone poles and utility boxes dominates a few major downtown streets, while on others, it has been placed underground. Certain downtowns have decorative lighting systems, benches, textured walkways, and other features aimed at encouraging people to linger, while others view loitering with a great deal of alarm. Street design and embellishments tend to accentuate differences in street morphology. Straight, wide streets with no vegetation can seem bleak indeed.

In spite of all the possible variations in downtown street design, the basic spatial arrangement of streets alone can be of great significance. In Manhattan, for example, there is a combination of street types with wide streets and short blocks going in one direction and narrow streets with long blocks running in the other, and this has had a major impact on downtown land uses. High-quality shopping and business addresses have traditionally been on the north-south avenues rather than the east-west streets. The avenues are not only wider and grander, but the blocks are much shorter. Walking along Fifth Avenue is fun in part because pedestrians feel that they are making rapid progress and because the frequently encountered street corners tend to be livelier than the middle sections of long blocks. In addition, the east-west blocks are shorter on the East Side of Manhattan and where semidiagonal Broadway intersects the grid on the West Side. The very best locations for business and residential uses tend to be where the blocks are small and the corners are numerous. The social geography of the central city has thus been shaped, in part, by street morphology.

When there are two or more competing grids, the contrasts in land uses are often extremely marked. The downtown is often "contained" in a single grid while adjacent grids house the downtown frame or support activities. In San Francisco, for example, a grid of small blocks and narrow streets exists north of Market Street, which is set off from it at a forty-five-degree angle. To the south of Market Street, there are long blocks and wider streets running parallel to that wide boulevard. For nearly a century, prime office activities and high-

Atlanta

Guide to the Letters

A—Arts or performing arts district; **C**—Civic space; **Cv**—Convention center; **E**—Education (college or university); **G**—Government (state capitol or city hall); **H**—Historic district; **L**—Library; **M**—Museum, art gallery; **Mx**—Mixed use complex; **P**—Park, civic open space; **R**—Retail and/or entertainment; **Rs**—Residential; **S**—Sports arena or stadium; **Z**—Zoo.

(Map credit) All sixteen maps are by Nina Veregge. They are intended for conceptual purposes and are not meant to be accurate portrayals of all streets or exact building locations.

Baltimore

Guide to the Letters

A—Arts or performing arts district; **C**—Civic space; **Cv**—Convention center; **E**—Education (college or university); **G**—Government (state capitol or city hall); **H**—Historic district; **L**—Library; **M**—Museum, art gallery; **Mx**—Mixed use complex; **P**—Park, civic open space; **R**—Retail and/or entertainment; **Rs**—Residential; **S**—Sports arena or stadium; **Z**—Zoo.

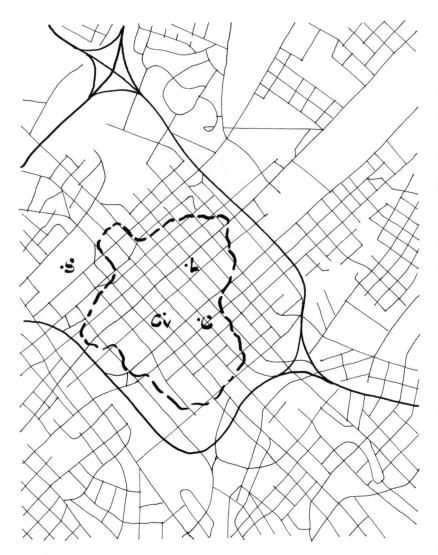

Charlotte

Guide to the Letters
A—Arts or performing arts district; **C**—Civic space; **Cv**—Convention center; **E**—Education (college or university); **G**—Government (state capitol or city hall); **H**—Historic district; **L**—Library; **M**—Museum, art gallery; **Mx**—Mixed use complex; **P**—Park, civic open space; **R**—Retail and/or entertainment; **Rs**—Residential; **S**—Sports arena or stadium; **Z**—Zoo.

Cleveland

Guide to the Letters

A—Arts or performing arts district; **C**—Civic space; **Cv**—Convention center; **E**—Education (college or university); **G**—Government (state capitol or city hall); **H**—Historic district; **L**—Library; **M**—Museum, art gallery; **Mx**—Mixed use complex; **P**—Park, civic open space; **R**—Retail and/or entertainment; **Rs**—Residential; **S**—Sports arena or stadium; **Z**—Zoo.

Columbus

Guide to the Letters

A—Arts or performing arts district; **C**—Civic space; **Cv**—Convention center; **E**—Education (college or university); **G**—Government (state capitol or city hall); **H**—Historic district; **L**—Library; **M**—Museum, art gallery; **Mx**—Mixed use complex; **P**—Park, civic open space; **R**—Retail and/or entertainment; **Rs**—Residential; **S**—Sports arena or stadium; **Z**—Zoo.

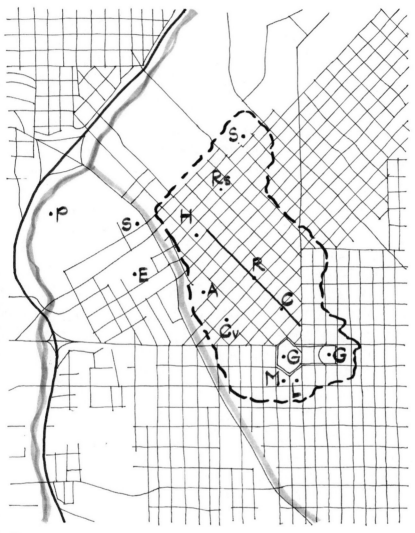

Denver

Guide to the Letters
A—Arts or performing arts district; **C**—Civic space; **Cv**—Convention center; **E**—Education (college or university); **G**—Government (state capitol or city hall); **H**—Historic district; **L**—Library; **M**—Museum, art gallery; **Mx**—Mixed use complex; **P**—Park, civic open space; **R**—Retail and/or entertainment; **Rs**—Residential; **S**—Sports arena or stadium; **Z**—Zoo.

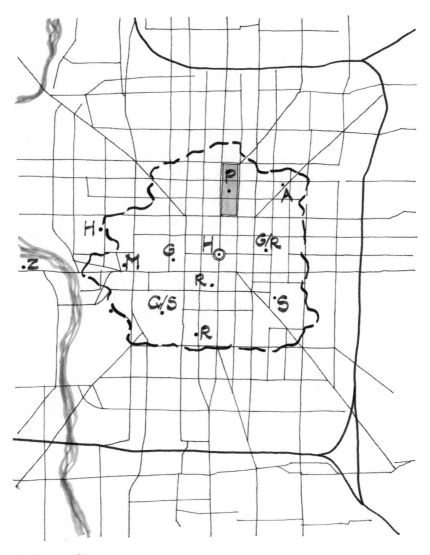

Indianapolis

Guide to the Letters

A—Arts or performing arts district; **C**—Civic space; **Cv**—Convention center; **E**—Education (college or university); **G**—Government (state capitol or city hall); **H**—Historic district; **L**—Library; **M**—Museum, art gallery; **Mx**—Mixed use complex; **P**—Park, civic open space; **R**—Retail and/or entertainment; **Rs**—Residential; **S**—Sports arena or stadium; **Z**—Zoo.

Minneapolis

Guide to the Letters

A—Arts or performing arts district; **C**—Civic space; **Cv**—Convention center; **E**—Education (college or university); **G**—Government (state capitol or city hall); **H**—Historic district; **L**—Library; **M**—Museum, art gallery; **Mx**—Mixed use complex; **P**—Park, civic open space; **R**—Retail and/or entertainment; **Rs**—Residential; **S**—Sports arena or stadium; **Z**—Zoo.

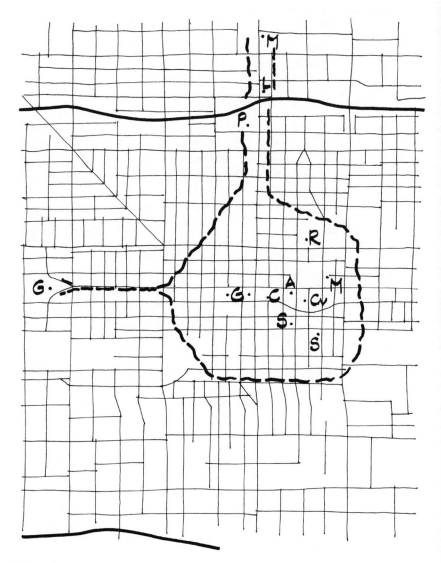

Phoenix

Guide to the Letters

A—Arts or performing arts district; **C**—Civic space; **Cv**—Convention center; **E**—Education (college or university); **G**—Government (state capitol or city hall); **H**—Historic district; **L**—Library; **M**—Museum, art gallery; **Mx**—Mixed use complex; **P**—Park, civic open space; **R**—Retail and/or entertainment; **Rs**—Residential; **S**—Sports arena or stadium; **Z**—Zoo.

Pittsburgh

Guide to the Letters

A—Arts or performing arts district; **C**—Civic space; **Cv**—Convention center; **E**—Education (college or university); **G**—Government (state capitol or city hall); **H**—Historic district; **L**—Library; **M**—Museum, art gallery; **Mx**—Mixed use complex; **P**—Park, civic open space; **R**—Retail and/or entertainment; **Rs**—Residential; **S**—Sports arena or stadium; **Z**—Zoo.

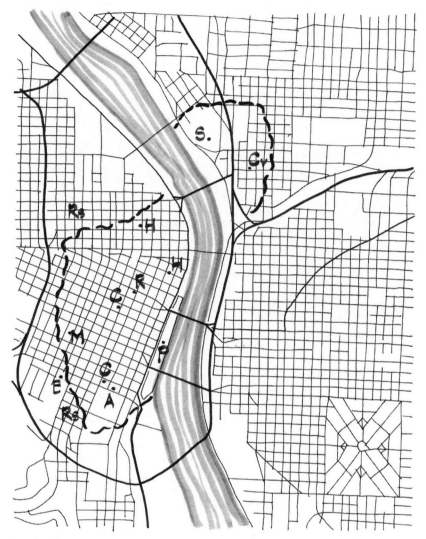

Portland

Guide to the Letters

A—Arts or performing arts district; **C**—Civic space; **Cv**—Convention center; **E**—Education (college or university); **G**—Government (state capitol or city hall); **H**—Historic district; **L**—Library; **M**—Museum, art gallery; **Mx**—Mixed use complex; **P**—Park, civic open space; **R**—Retail and/or entertainment; **Rs**—Residential; **S**—Sports arena or stadium; **Z**—Zoo.

Providence

Guide to the Letters
A—Arts or performing arts district; **C**—Civic space; **Cv**—Convention center; **E**—Education (college or university); **G**—Government (state capitol or city hall); **H**—Historic district; **L**—Library; **M**—Museum, art gallery; **Mx**—Mixed use complex; **P**—Park, civic open space; **R**—Retail and/or entertainment; **Rs**—Residential; **S**—Sports arena or stadium; **Z**—Zoo.

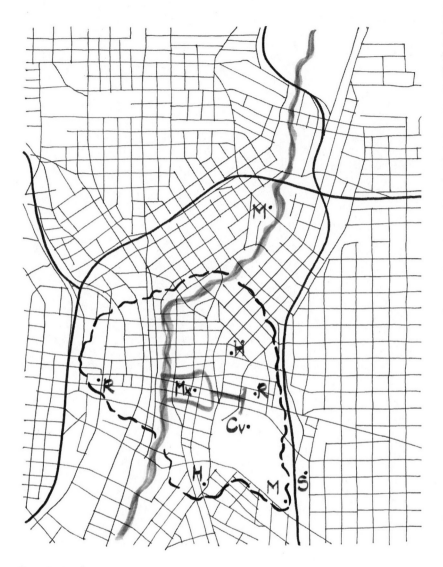

San Antonio

Guide to the Letters

A—Arts or performing arts district; **C**—Civic space; **Cv**—Convention center; **E**—Education (college or university); **G**—Government (state capitol or city hall); **H**—Historic district; **L**—Library; **M**—Museum, art gallery; **Mx**—Mixed use complex; **P**—Park, civic open space; **R**—Retail and/or entertainment; **Rs**—Residential; **S**—Sports arena or stadium; **Z**—Zoo.

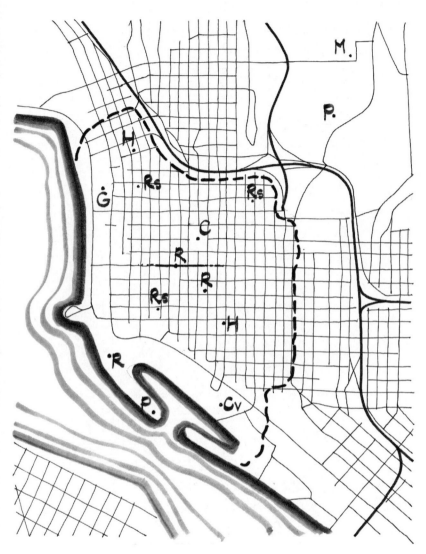

San Diego

Guide to the Letters
A—Arts or performing arts district; **C**—Civic space; **Cv**—Convention center; **E**—Education (college or university); **G**—Government (state capitol or city hall); **H**—Historic district; **L**—Library; **M**—Museum, art gallery; **Mx**—Mixed use complex; **P**—Park, civic open space; **R**—Retail and/or entertainment; **Rs**—Residential; **S**—Sports arena or stadium; **Z**—Zoo.

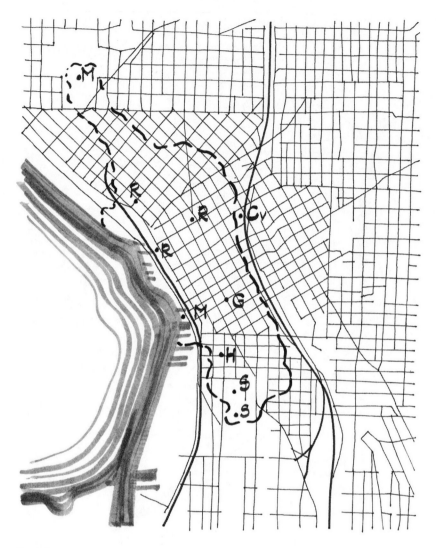

Seattle

Guide to the Letters

A—Arts or performing arts district; C—Civic space; Cv—Convention center; E—Education (college or university); G—Government (state capitol or city hall); H—Historic district; L—Library; M—Museum, art gallery; Mx—Mixed use complex; P—Park, civic open space; R—Retail and/or entertainment; Rs—Residential; S—Sports arena or stadium; Z—Zoo.

St. Louis

Guide to the Letters
A—Arts or performing arts district; **C**—Civic space; **Cv**—Convention center; **E**—
Education (college or university); **G**—Government (state capitol or city hall); **H**—
Historic district; **L**—Library; **M**—Museum, art gallery; **Mx**—Mixed use complex;
P—Park, civic open space; **R**—Retail and/or entertainment; **Rs**—Residential; **S**—
Sports arena or stadium; **Z**—Zoo.

end shopping remained almost exclusively north of Market. The scale of the city was cozier on the north side, and the combination of the width of Market Street, the awkward angles where the two grids came together, and the long blocks to the south meant that downtown became bifurcated into two very separate districts. From Market Street, there are lots of paths to the north but few to the south. In recent years, a combination of massive urban renewal schemes and relatively liberal zoning south of Market has gradually diminished the polarity, but the two areas are still quite different. Mismatched grids can be as important as physical barriers in shaping downtowns.

Differing grids have also made for awkward linkages in other downtowns. In Pittsburgh, for example, the main shopping and office district has traditionally been located on one grid while the separate grid north of Liberty Street has languished. The latter is now the focus of a historic preservation area known as the Cultural District because its buildings date mostly from before 1910. In Atlanta, the northward march of the central business district has in part been due to the confusion associated with various grids coming together at Five Points in the southern part of the downtown. In recent years, high-order activities have settled onto Peachtree Street and its associated relatively legible grid. In Denver, the monumental state capitol and civic complex are located on a separate grid from the downtown core. As a result, there is an awkward approach to them and they have not been as fully integrated into the downtown as they might have been. Downtown Minneapolis has also had problems breaking out of its one dominant grid. Although breaks in the urban fabric caused by varying grids may be easily crossed by those determined to do so, for those just strolling and following a path of least resistance, awkward intersections can be a significant barrier. Stores and restaurants can succeed or fail based on subtle differences in perceived accessibility.

Both street width and block size have also played roles in the spatial organization of a variety of cities. In Salt Lake City, the streets and blocks were laid out initially to be inordinately large. Before long, however, local leaders realized that the huge alleyless blocks were too large to accommodate easily serviceable buildings, and so smaller streets and alleys were cut through. Even today, however, parts of downtown Salt Lake City seem more akin to a suburban office park than to a traditional downtown, in that lots are large enough to accommodate new "skyscraper in a park" designs complete with large lawns and surface parking. Its grid of wide streets makes for easy traffic flow and a sense of spaciousness. On the other hand, there is a relatively weak sense of enclo-

sure or "urbanity." Street morphology plays an important role in differentiating Salt Lake City from, say, Portland or Pittsburgh.

Downtown Portland represents the polar opposite of Salt Lake City in terms of the scale of its street pattern. Portland's downtown, which is mostly located on one grid, consists of narrow streets and small blocks. Compared to Salt Lake City, it seems dense and even claustrophobic, especially since the streets are lined with trees and there are few vacant lots. But Portland has become the model that most downtown planners want to emulate. It is so compact that no destination can be very far away from any other. In addition, required street-level retailing, a light rail system, and two major streets given over to bus malls make getting around downtown easy and pleasant. The existence of one dominant downtown grid, coupled with the river on one side and hills on the other acting as visual boundaries, makes downtown Portland extremely legible or understandable at a glance. Consequently, people seem more willing to walk and explore new attractions in and around downtown. It works very well as a pedestrian city. Mixed grids and confusing street patterns can have the opposite effect. But scale alone is also important. A grid of large blocks, wide streets, and buildings set back from the street can intimidate pedestrians and dissuade them from exploring.

Exceptions to the Grid: Medieval, Modernist, and Baroque Touches

With the possible exception of Boston, there are no major American downtowns with squiggly medieval street patterns. Even those that started out that way have largely been modified into a sort of semi-grid. Indeed, meandering medieval-style streets were so rare in America that when they were introduced in cemeteries and suburbs in the nineteenth century, they were viewed as major curiosities and even tourist attractions. The lack of irregular streets and blocks has had an enormous impact on the size, shape, and character of typical downtown architecture. In the "medieval" cities of Europe, buildings come in all shapes and sizes, from tiny shops to huge palaces. Some have rounded corners and others have anywhere from three to seven sides. Since the buildings and streets gradually evolved together, the morphology of the central city reflects continuous give and take, compromise and adjustment. Sometimes buildings encroached on the street and sometimes streets were widened. While a few hints of this pattern can still be found in Boston, where rounded facades and unusually shaped buildings still exist, the typical American downtown features building footprints that are far less interesting.

In American cities, the prevailing grid shaped much of the architecture. Since most American downtowns were laid out as speculative ventures, the grid was preferred for its simplicity. Rectangular blocks meant easily marketable rectangular lots with immediately recognizable dimensions and locations. A twenty-five- by one-hundred-foot corner lot was a recognizable commodity that could be sold to far-away buyers. As a result, the built environment quickly became recognizable and predictable as well. Look-alike buildings with twenty-five- or fifty-foot facades became the norm in most downtowns. As Victorian architecture with its more varied facades came into vogue in the late 1800s, the cityscape became a bit more picturesque, but building variety was still constrained by morphology. As individual enterprises sought increased architectural identity, building height and facade detailing were the only possible variations. Many downtowns began to look a bit jumbled as ten-story Italianate buildings appeared next to two-story Federal-style structures. The footprints, backs, and sides of typical buildings changed little, if at all.

As the twentieth century progressed, other challenges arose which resulted from problematic relationships between buildings and the grid. For a while, buildings simply became larger as wealthy owners assembled adjacent lots. When modernist planning ideas arrived on the scene during the 1950s and 1960s, many cities enacted codes and regulations that encouraged lot assembly for open plazas as well. Where urban renewal projects played a role in clearing land, huge setbacks and (still rectangular) superblocks were sometimes created. No longer did streets and buildings interact. Streets often became irrelevant as building facades avoided any contact with them. As if to increase variety, buildings were occasionally set at odd angles to the street, sometimes contributing to an awkward tension between the two. In the process, the grid lost some of its power over downtown morphology, but sometimes the price was high, as the role of the street diminished both visually and functionally.

There are a few examples of American downtowns with grand, Baroque-style radial streets, although most of them have been designed in a sort of ad hoc, after-the-fact manner. Fewer still have the traffic circles and roundabouts associated with Baroque planning. By far the best example is Washington, D.C., although Indianapolis, Detroit, and Philadelphia have at least a few traits. Pioneered in European capitals such as Paris, radial streets can add both monumentality and confusion to the downtown experience. Since proper radial streets require some point to radiate from, they are usually built outward from a central point. Indianapolis, for example, has Monument Circle at its center,

The Charles Center superblock in Baltimore

although its remaining radial streets do not start there. Similarly, Philadelphia has a monumental central square and city hall with a radial grand boulevard leading from it to the art museum in Fairmount Park. Detroit has two radial boulevards, although its central meeting place is no longer as important as it once was.

In several other cities, there are accidental radial streets that have resulted from varying grids and/or antecedent roads. In New York City, for example, Broadway has created odd-sized blocks where it has crossed the prevailing grid. The famous "prow" of the three-sided Flatiron Building resulted from the use of a small, triangular lot.

The advantage of radial streets and mismatched grids is that they can provide a sense of closure and perspective in what might otherwise be endless, straight streets. When there are buildings at the end of the street, there is someplace to head for as you walk. Such sites also provide the opportunity for purposeful monumentality, as when the Colorado State Capitol looms at the end of Denver's major shopping street as a result of its being located on a different grid. There are, of course, other types of grid interruptions. Chief among them are parks, squares, and superblocks such as those in Savannah or Phoenix. I will

Monument Circle: the symbolic center of Indianapolis

discuss these in the following chapter under the heading of public and civic
space.

Another Baroque element that is nearly absent from the American downtown is the *ringstrasse* or ring road in the form of a grand boulevard, although
there are sometimes minimalist variations on the theme. In its classic form, as
found in cities such as Vienna and Moscow, one or more monumental ring
roads fill the space once occupied by city walls and related fortifications. The
grandeur of the ring road thus contrasts with the tighter spaces of the earlier
medieval city inside the boulevard. Lined with trees, opera houses, palaces, and
large apartment buildings, the ring road can add elegance to cities once characterized mainly by narrow lanes and small market squares. While "ring" roads
imply that the grand boulevard surrounds the old city, in many cases, there was
only partial enclosure.

No American downtowns have complete, Viennese-style ring roads. There are, however, many examples of relatively grand parkways on the edges of older downtown grids. Many of them date from the City Beautiful Movement of the early twentieth century, but some were built before or after that era. In Cincinnati, for example, an old canal was filled during the early 1900s and replaced with a tree-lined parkway that separated the old downtown core from the lower-density frame. A few upscale apartment buildings were constructed along the parkway that still exist today. It's nice, but it's no Vienna. Similar "edge of downtown" parkways were built in Indianapolis, Milwaukee, Richmond, and Sacramento. In some cases, waterfront parkways were added to increase the grandeur of the city. Harbor Drive in San Diego and Marconi Boulevard in Columbus are examples. The primary goal in nearly every case, however, was to facilitate traffic flows, although aesthetics usually played a minor role. Since many parkways and boulevards were completed either just before or during the Depression, architectural grandeur came slowly. By the time prosperity returned in the 1950s, new types of highways were all the rage.

Freeways and Downtown Definition

It is interesting to speculate what American downtowns would be like today if the tradition of *ringstrasse*-like parkways had been expanded instead of freeway inner belts. While grand boulevards tie the central core to surrounding neighborhoods by adding amenities such as parks and museums, expressways have cut off the downtown with a wall of isolating highways and ramps. There are relatively few routes into downtowns today, since most are surrounded with inner belt highways that are breached in only a limited number of places. On the other hand, the combination of elevated highways and bridges can make for some pretty spectacular views of downtown that help to make it a monumental place, at least from a distance.

But the negatives far outweigh the positives. The inner belt "wall" is not the only thing that isolates downtown from other parts of the central city. The fact that most people enter downtowns via freeways means that contact with the rest of the central city is minimal. It is a chicken-and-egg situation. Since the commercial strips leading in and out of downtowns grew up depending on business from first a trolley-riding and later an auto-using clientele, most have declined precipitously as those clienteles have been taken away. As a result, the "inner city" looks decrepit and scary and so downtown workers tend to avoid it. The inner belt "wall" may even be viewed positively as a partial defense

against the potentially unruly neighborhoods just beyond. When inner belts are combined with rivers, bridges, and hills, the downtown might be seen as occupying a classic defensible site, an island of glitter in a sea of decline.

Problems may occur when contact between the downtown core and nearby neighborhoods is problematic. In Atlanta, for example, an awkward tangle of freeways separates the downtown core from the booming Midtown area just to the north. What should be one big downtown has been cut in two. In Seattle and Columbus, freeways separate downtown from gentrified neighborhoods close by, although Seattle has constructed a "freeway park" over the interstate to partially ameliorate the problem. In Cleveland and St. Louis, expressways separate the downtown cores from their waterfronts. Could more aesthetic and pedestrian-friendly grand boulevards have served the purpose of moving traffic and linking downtown to its environment better?

For better or for worse, freeways often define American downtowns as distinctive places. A downtown defined by an inner belt usually has an abundance of space that is expected to be eventually redeveloped for "higher and better" uses. As mentioned in chapter 2, however, there is often too much land, and so space-extensive uses, such as stadiums, must be built to fill it. Still, as in the walled cities of old, inner belts help people to visualize where downtown begins. Occasionally, cities have used topography to heighten the visual excitement of arriving in downtown. Pittsburgh and San Diego provide excellent examples of this. In the former, the visitor speeds (except at rush hour) along a rustic freeway only to enter a tunnel and emerge with a full, close-up view of the city skyline. In San Diego, one can travel by freeway through a green canyon in Balboa Park to the very edge of the downtown core. In these cases, there is something to be said for the view from the road.

Auto Dependency and the Lack of Great Streets

There is some irony in the fact that even though we have become more dependent on movement through urban space, we seem to have lost our ability to create and maintain truly great streets.[5] Most of our modern downtowns consist of a combination of freeways and relatively nondescript commercial thoroughfares. The term *great street* is, of course, not always easy to define, but there are many examples worldwide of possible contenders. The most obvious examples are grand boulevards such as the Champs Élysées in Paris, the Diagonal in Barcelona, or the Ringstrasse in Vienna, but there are many smaller great streets as well. North American examples might include Fifth Avenue in

New York City, Commonwealth Avenue in Boston, Michigan Avenue in Chicago, Market Street in San Francisco, and Wilshire Boulevard in Los Angeles. Most of these streets have been well known for at least a century, although a few, such as Wilshire, Peachtree Street in Atlanta, and perhaps Central Avenue in Phoenix have gained fame only in the last fifty years. Still, there do not seem to be many new great streets on the horizon. Downtown streets are increasingly geared for smooth, one-way traffic rather than pomp and spectacle. In suburban edge cities, the prospects for greatness are even more dismal among the eight-lane highways fronted by vast parking lots and banal signage. Still, attempts to create personable, memorable, and even beautiful great streets in our cities are worth monitoring as part of the quest to create good downtowns.

And now I use the rating system described above to examine street morphology in the sixteen downtown case studies.

Variable 2. Street Morphology: Great Streets, Small Blocks, and Connectivity

Portland: A (10)

Portland is a textbook example of a city core with a clear-cut grid, small blocks and streets, and a well-defined separate grid for its historic district. There are also distinctive "major streets" for bus malls and light rail. It has all the things that current writers define as good urban features for scale and connectivity. Since the riverfront highway was taken down, intrusive highways have been limited to the below-grade inner belt on the west and south and two somewhat unaesthetic bridge ramps. While there are no truly "great streets" or grand boulevards, there is good use of street trees, bus shelters, and statuary and other "street art." However, if the secondary or overflow district across the river is included in the definition of downtown Portland, the overall design must be seen as less successful. Freeways and bland highways chop up the secondary district. The Willamette River, a major amenity, acts as a barrier to downtown expansion and coherence.

San Diego: A− (9)

San Diego has a simple downtown grid with small blocks and narrow streets but has failed to embellish them to the same degree as Portland. The inner belt on the north and east separates downtown from Balboa Park and the Uptown district. There are some almost "great streets" (Broadway and Harbor Drive)

Small blocks and pedestrian-friendly streets in Portland

but they are not quite there yet, although they do have distinctive personalities. Fifth Avenue in the Gaslamp Quarter provides a cozy setting for restaurants and nightlife. Busy streets such as Pacific Highway separate the downtown core from the waterfront. Large superblocks for the civic center, Horton Plaza shopping center, the convention center, and some housing developments disrupt the grid and form intra-downtown barriers in several places. Flows to neighborhoods near downtown are good, although there are some ugly freeway ramps to the east.

Indianapolis: A− (9)

Indianapolis has a combination of grid and radial boulevards in the manner of Washington, D.C., although some of the radial streets have been partly obliterated by superblocks. There is a consensus node at Monument Circle, which has historically acted as a focus for urban identity. Inner belt freeways are located well beyond the downtown fringe, although wide and busy surface streets do sometimes act as comparable barriers. Elevated railroad tracks just to the south of the downtown core act as a major barrier to southern support zones but do help to keep the core compact. There are some pretty good "main streets" with concentrated activity but no really great streets. Enclosed views of

the state capitol and Monument Circle add personality to the street system. Flows outward to the neighborhoods are generally good, but there are big gaps.

Columbus: B+ (8)

Downtown Columbus features a grid of intermediate-sized streets and blocks of mixed sizes with only a few interruptions and variations. Superblocks have cut off some fringe areas, however. There is a consensus "main street" (High) which has historically been lined with shops and offices and a reasonably successful "great" street (Broad) which is extra wide and makes a grand entrance into the core over the Scioto River. Marconi Boulevard/Civic Center Drive follows the riverfront through downtown and provides visual access to it, as well as a setting for festivals and special events. However, many streets are underutilized due to parking lots and other gaps in the urban fabric. Flows to several nearby neighborhoods are good, although the close-in but below-grade inner belt makes for some awkward transitions.

Denver: B (7)

Denver's core consists of two major grid systems and, as a result, there are many awkward intersections and some intra-core confusion, although occasional street-end views of major buildings make for visual interest. Within districts as defined by grids, however, there is little confusion. With its successful bus mall and landscaped retail-recreational ambience, Sixteenth Street acts as a notable "great" street surrounded by a supporting grid. Superblocks for government offices, convention center, performing arts center, and educational complexes ring the downtown core and impede access to nearby districts— sometimes in conjunction with mixed grids. There is no close-in beltway and so freeway barriers are minimal. However, railroad tracks and a vast and semi-empty area to the northwest of the core make for poor linkages to the neighborhoods beyond.

Seattle: B (7)

The core of downtown Seattle is located on a clear-cut grid with varied street widths and block sizes, but there are some messy transitions since there are four different adjacent grid systems. Long north-south blocks and steep east-west topography make for occasionally difficult pedestrian access, but the views down streets can be extraordinary. The inner belt to the east and the elevated highway along the waterfront give the downtown some personality but also

The Sixteenth Street Mall, a "great street" in Denver

some noisy tension and a sense of separation between the water and the downtown core. Freeway Park, a platform over the below-grade highway helps connect downtown to nearby neighborhoods but only in a small section, while Westlake Park helps to soften an awkward grid transition. Historic Pioneer Square occupies a separate grid with small streets and blocks and a strong sense of identity. There are no "great" streets.

Minneapolis: B (7)

Downtown Minneapolis has one dominant grid with reasonable-sized blocks and streets, although there are some awkward transitions to the south and west. There is an excellent attempt at the development of a consensus "main street" in the Nicollet Avenue bus mall and shopping street. The Third Avenue Bridge makes for a grand "city entry" with a river and skyline view.

Overhead skyways in Minneapolis block the views but help to connect downtown buildings.

However, spaghetti-like tangles of inner belts add barriers, especially to the west, where they are combined with a wall of massive parking garages. Superblocks are appropriately located at the edge of the core, such as at the convention center and Humphrey Dome. There are a few messy transitions for pedestrians such as around the Walker Art Center. Many otherwise impressive street vistas are hurt by skyways that clutter any sense of perspective.

St. Louis: B (7)

Downtown St. Louis has nice small streets and blocks in a small core area, but there is almost a sense of congestion and claustrophobia where tall buildings enclose tiny lanes. A freeway and bridge ramps form a wall along the southern edge of downtown and a waterfront highway separates the downtown core from Laclede's Landing and the Arch. Beyond the core, a number of broad east-west streets lead out of town, suggesting movement rather than intra-core interaction. There have been attempts to create grand streets with monumental vistas, but most seem underutilized. Large complexes such as sports arenas and convention centers act as superblocks, which disrupt the grid. Still, the Arch makes for a grand sense of perspective as one enters the city.

Baltimore: B− (6)

Downtown Baltimore features a grid of small blocks and streets surrounded by an even smaller grid (due to core rebuilding after a nineteenth-century fire). A wide and busy highway along the waterfront separates the core from Harborplace and other recreational activities and there is an inner belt barrier along much of the eastern border. Superblocks resulting from urban renewal projects such as Charles Center also disrupt the grid, and a maze of freeways to the south makes for considerable separation and confusion around the sports complexes. There is a traditional "main street" (Charles) but it seems increasingly underutilized, as much of the life of the city is moving eastward along the waterfront. There are several mismatched grids around the CBD edge, but the core is generally legible.

Charles Street, the "main street" of downtown Baltimore

Charlotte: B− (6)

Charlotte has a consensus "main street" (Tryon) with impressive buildings and some good embellishments (art), along with a reasonably clear-cut grid. The sense of downtown trails off rapidly, however, in all directions. Super-blocks for convention and government complexes, a football stadium, industry, and public housing disrupt the grid. There are many gaps in the urban fabric, especially around the edges. Mismatched grids and highways make for some awkward transitions where the CBD grid meets the surrounding neighborhoods, but the internal grid is legible and spacious.

Providence: B− (6)

Downtown Providence is located on a very small-scale semigrid with tiny lanes. Gently curving streets and small blocks make for an almost medieval urban ambience compared to most North American cities. The small size of the downtown core and the distinct boundaries between it and the areas "outside" is reminiscent of a walled city. Linkages are very good within the core but somewhat awkward at the edges due to rivers, topography, and freeways (and, to a lesser extent, the capitol grounds). There are no grand or "great" streets although many provide interesting perspectives. The main problem is the downtown grid is too small and constraining.

Phoenix: B− (6)

The Phoenix grid extends well beyond the CBD core, and that is both a negative and positive characteristic. Flows in and out of the core are relatively easy but there is little sense of confinement or enclosure (the opposite of Providence). Like the metropolitan area, the downtown sprawls, with the state capitol and art museum located many blocks away in different directions. Only the railroad tracks limit the expansion of the downtown core to the south. The streets are generally wide and bland with little differentiation. Central Avenue seems intended to be a great street but acts to drain rather than reinforce the core as activities have spread northward. There are many large superblock barriers (ballpark, arena, convention center, Arizona Center, etc.), which disrupt flows. On-street parking is very limited and so most streets act as throughways rather than settings for downtown activity.

Central Avenue in Phoenix acts as a spine leading out of downtown.

San Antonio: B− (5)

San Antonio's downtown features a mismatched and irregular grid with individual sections attempting to relate to the river in different ways. There are modest east-west commercial streets, but since most activity is focused on the river, they lack a sense of grandeur. Attempts are being made to create pedestrian pathways to tie the core together and increase coherence. Although the slightly confusing relationship of the Riverwalk to the street system often adds a pleasant sense of mystery, it detracts from street vitality. There are awkward linkages to areas beyond the core due to different street grids and freeway barriers. Streets play second fiddle to the river below grade but contribute to the "cozy" and historic feel of the city.

Pittsburgh: C+ (5)

Pittsburgh combines two separate grids in its small and compact downtown—one with small blocks and the other (once skid row and now the Cultural District) with longer, larger blocks. There are some awkward transitions, but the confusion doesn't last long as end-point landmarks quickly come into view. Highways line most of the waterfront, precluding streets from contributing to

the riverfront site, although grand bridge entrances can be exciting. Access to nearby neighborhoods is often very difficult due to topography and massive freeways, ramps, and awkward street systems. There are no "great streets" although Grant makes an attempt at it for a couple of blocks.

Cleveland: C+ (5)

Squeezed between Lake Erie and the meandering Cuyahoga River to the west and south, downtown Cleveland would be somewhat isolated even with the best street system. As it is, mismatched grids with streets fanning out to the east contribute to the confusion. This mixed grid, coupled with constrained physical site, makes for some awkward linkages. Public Square provides a sense of focus, which does help hold things together. Euclid Avenue was once a "great street" but it has fallen on hard times over the past fifty years. Freeways and bridges are too dominant, although they sometimes provide visual interest and grand entries into the city. Access to the lakefront is poor but improving.

Atlanta: C (4)

Atlanta's downtown grew up around a variety of mismatched grids, each relating to rail lines and topography in different ways. Today railroad tracks and highways cut through the core at awkward angles, and layers of transportation in and around a topographic "gulch" contribute to the confusion. Peachtree Street aspires to "great street" status but it is more successful once it reaches the Midtown area. Huge, superblock developments such as the Georgia Dome add barriers to interaction with surrounding neighborhoods. All of this makes downtown Atlanta less cohesive and legible than most comparable centers, and so many activities cling to the linear Peachtree corridor.

Summary: Street Morphology

> Portland 10
> San Diego 9
> Indianapolis 9
> Columbus 8
> Denver 7
> Seattle 7
> Minneapolis 7
> St. Louis 7
> Baltimore 6

Charlotte 6
Providence 6
Phoenix 6
San Antonio 5
Pittsburgh 5
Cleveland 5
Atlanta 4

Public Space and Civic Architecture

Of all the issues surrounding the design and revitalization of American downtowns, perhaps none is more controversial than the role of access to good public space. A growing literature has emerged alleging that the public city of parks, plazas, and streets has been marginalized or even eliminated in favor of private and semiprivate corporate plazas, shopping malls, and other controlled, interior spaces.[6] The basic assertion is that "for profit" commercial spaces are now so attractive and ubiquitous that they have become the de facto public spaces of the city. Competition from existing, truly public settings is all but impossible. As public space has become peripheral to the activity patterns of downtown, so too have the people who use them. Only those who can afford to shop in mall atria and dine in "garden courts" can participate in the "public" life of the city. Hidden cameras and security patrols make sure that the disheveled masses do not become at home in the new corporate plazas. The homeless and downtrodden are left to fend for themselves in poorly maintained parks and plazas. Indeed, the term *street* (as in street person) has joined the ranks of denigrating phrases along with earlier place-based descriptors such as *alley rat* or *dead-end kid*.

Although the above assertions are common and widely accepted, they have not been adequately tested in a variety of settings. Rather, generalizations and even "theories" have been derived from a few famous (infamous) examples in overrepresented cities such as Los Angeles, New York, and Detroit. In Los Angeles, for example, the shiny, new corporate plazas are primarily located high up on Bunker Hill while Pershing Square, the traditional public plaza, is "down" in the less-affluent, Latino part of the downtown. There is little cross-fertilization. While segregation is not necessarily strictly enforced, it is easy to conclude that the "good" part of town has been privatized.

In New York City, some highly visible and accessible public settings, such as

Bryant Park on Forty-second Street, have been cleaned up and sanitized after years of having reputations as notorious centers of drug dealing and danger. While it is hard to argue against making places clean and safe, there are concerns that the "new and safer" versions of public parks demand a degree of exclusivity that makes them less than truly public. Anyway, many people prefer the glass-enclosed plazas found in many new Midtown corporate centers.

Beyond these few well-known examples, there is little evidence either to support or refute the alleged trend toward the privatization of open space. We do not have enough solid, empirical evidence from a large sample of American downtowns to agree or disagree on the issue of "the end of public space." We do not even know exactly what kinds of limits on behavior, if any, should define public versus private spaces. It was not that long ago, for example, that people could be arrested for "loitering" or "unruly" behavior in public spaces. Often, members of minority groups and women were excluded from public spaces at certain times. It is possible that part of the enthusiasm for semipublic or even privatized spaces is due to diminished enforcement of civil behavior in public spaces. Of course, there is the argument that the two trends go hand in hand. That is, since the middle class can escape to pleasant, privatized space, there is no longer any reason to maintain or even enforce laws in the increasingly marginal public realm.

The Evolution of Public Space in American Downtowns

Other than narrow, winding streets, medieval cities had little or no public space. The idea of having parks or "open space" in tiny, fortified urban areas in which no place was very far from the countryside seemed nonsensical. Blank walls predominated and even cathedrals were normally embedded in the dense, built environment. By the Renaissance, however, cities began to open up and a sense of civic identity began to take shape. In cities from Brugge in the north to Siena in the south, grand city halls bristling with monumental towers brought pride and identity to thriving, prosperous places. Today, the classic, if Eurocentric, image of the ideal city typically involves some combination of a grand open plaza and monumental civic, or perhaps religious, architecture. The plaza major in Latin American cities epitomizes this idealized image. Such plazas define the center of the city and are flanked by the cathedral, the cabildo (city hall), and the most important hotels, offices, and residences in the area.

Although plaza design and spatial organization are less predictable in Euro-

pean cities, the prevalence and importance of piazzas, churches, and palaces give most cities there a strong sense of place and identity. In addition, streets themselves are important public spaces. In compact, preindustrial cities, most people walk to their destinations and much of the life and conversation in every neighborhood takes place on streets. Even when blank walls predominate, as they do in parts of Mediterranean Europe and Latin America, streets can be lively.

Grand public spaces were not always a high priority during the initial planning of American cities. About the only green spaces in compact cities such as Boston and New York were cemeteries. Sometimes, land originally designated for parks and plazas was built over as land values increased and the demand for space became intense. Other open spaces were left to languish as the core of the city moved away from the jumble of the commercial waterfront or, later, railroad yards. Even Jackson Square, the now-revived historic center of New Orleans, became peripheral for a while as the downtown core moved across Canal Street. Still, good public open space was not absent in the typical American downtown. City Hall Park in New York City, Boston Common, and the Mall in Washington, D.C., were all laid out before 1850. In addition, large public parks such as Central Park in New York, Fairmount Park in Philadelphia, and Balboa Park in San Diego were established close to, but not in, the emerging central business districts.

During the late 1800s, many new state capitol buildings and towering county courthouses were constructed and set in spacious grounds. Most of these, however, were located on the periphery of the downtown core. State capitols were often located on hills overlooking business districts, as in Hartford, Providence, St. Paul, or Salt Lake City. Most downtowns were thought to be places for business, and not much attention was paid to designing pleasant spaces for enjoyment and relaxation. Perhaps the first "golden age" of civic open space resulted from the City Beautiful Movement around the turn of the twentieth century. It was then that the idea of a "civic center" with grand public buildings of a variety of kinds was born.[7]

The "White City" of classical-style buildings set in monumental open spaces at the Chicago World's Fair in 1893 inspired many American cities to dress up their downtowns with new parks, plazas, statues, fountains, and public buildings. Most American downtowns had been quickly thrown together over just a few decades and, compared to the older cities of Europe and Asia, they tended to lack grandeur. During the first fifteen years of the twentieth century, many

downtowns were opened up and embellished. Chicago created a vast, green park on the downtown lakefront, and civic "malls" were constructed in cities as diverse as Cleveland, St. Louis, San Francisco, and Denver. The trend continued with war memorials and other monuments after World War I and on into the Depression-era projects of the 1930s. By 1940, a number of additional cities such as Columbus, Indianapolis, San Diego, and New Orleans had new civic centers and grand public spaces. Ironically, shortly after many of the new "public" spaces were created, the Great Depression brought bread lines and hobo camps to central cities. Many public spaces, from Scollay Square in Boston to Pioneer Square in Seattle, became identified with "skid row." Public space gradually became space to be avoided, and the issue of "whose space is it?" arose in planning circles.

In a few cities, such as St. Louis, some really grand plans were begun during this period. Beginning in the 1930s and continuing into the 1970s, the old waterfront of St. Louis was cleared for a huge (national) park and the Gateway Arch. Many acres were planted with grass and trees. As the new park joined the City Beautiful plazas of downtown St. Louis, the downtown core came to have more green open space than most major American cities. While tourists have flocked to the attraction, it is not a foregone conclusion that the creation of huge amounts of grand and green civic space has played a role in stemming central city decline. Too much space can make a downtown feel empty. Milwaukee also created a vast, green park between the downtown core and Lake Michigan but it has only gradually figured out how to make the space less passive and more a part of downtown activity.

In spite of a few grand projects, as recently as 1960 parks and plazas were still not common in most American downtowns. Commercial streets were just that, and open, public spaces were few and far between. As urban renewal projects came to fruition in the late 1950s and early 1960s, the idea of open space in the form of "skyscrapers in a park" gained popularity. Streets, now thought to be dangerous intrusions into the urban core, were slated for elimination in favor of superblocks and even raised platforms with gardens and office towers. Although touted as public space, many of the new plazas were located so as to be remote from any possible "teeming" areas where open space was needed. In addition, reliance on extreme modernist design often made for vast, empty, windswept plazas devoid of either public life or commercial activity. Benches and other amenities that might invite people (especially poor people) to linger and hang out were purposefully omitted in many of the new plazas.

Reclaiming the Waterfront for Public Space

The most exciting opportunity for creating good public open space has typically occurred as deindustrialization freed up large amounts of land along rivers and harbors. It was clear by the late 1960s that the old port facilities and warehouses close to typical downtowns were no longer competitive, as newer port facilities were built downstream and truck-oriented distribution centers emerged along freeways. Cities such as Portland, Baltimore, San Diego, and San Francisco more than doubled the size of downtown park and plaza space through the recapturing of formerly industrial land. Typically, these open spaces became more central to the downtown core as office towers and hotels moved toward the new amenities.

In part, the successful integration of the waterfront parks with the city core was due to the fact that most cities had learned from past mistakes. Instead of constructing large, passive open areas, the newer projects usually included housing and commercial attractions. Portland's Tom McCall Waterfront Park is a case in point. Anchored by shops, restaurants, apartments, and a marina at one end and close-by brewpubs in the Yamhill Historic District at the other, the long, thin park seldom feels remote from the life of downtown. A monumental fountain adds a focus to the middle of the space. In addition, a large number of festivals and events have been initiated to make sure the park is familiar and used. A mixture of passive open space and intense commercial activity also characterizes the Baltimore Inner Harbor. Green parks and paved promenades share space with festival marketplaces and major attractions such as museums and an aquarium. Getting the mix right is often a challenge.

Corporate Plazas and Public-Private Partnerships

The glory days of urban renewal have passed. Using huge quantities of public money to clear vast tracts of urban space may or may not have been a mistake, depending on the project, but it will no doubt occur less often in the future. The combination of fiscal retrenchment, political opposition, and design options has led cities to the realization that the most feasible open space projects are those that combine private-sector investment with incentive zoning or some type of public subsidy. The idea is to cajole, persuade, and perhaps bribe developers to include "good" public space in their projects. This can be done through a variety of FAR (floor area ratio) requirements mandating that a certain percentage of the lot be left open, zoning bonuses for public amenities, and

Boat rides, parks, and museums on the Baltimore harbor

policies such as "one percent for art" that encourage sculpture gardens or public fountains. The results have been very mixed. Some "corporate plazas" have not worked at all, while others have worked so well that they have helped to marginalize "real" public space and contribute to social and economic segregation and polarization. This is especially true when real public parks and plazas are poorly maintained, occupy peripheral locations, and are left to the poor and downtrodden. Cities may thus have two types of open space—one for white-collar workers and well-heeled shoppers and the other for the less-affluent masses. Designers argue that the best way to revitalize the downtown is to make sure that it is a comfortable place for the middle class, and that requires "keeping the bums away."

Testing the Theories: Has Public Space Declined?

Although there has been a great deal of discussion and debate over "the end of public space," it is not an easy thing to measure objectively. Most authors, or so it seems, start out with a point of view and provide selective examples to prove their point. There are plenty of examples to demonstrate the validity of either side in most matters. For example, Los Angeles, often depicted as the

epitome of the inhospitable, postmodern fortress city, works pretty well as an example of the evildoings going on in the name of urban design.[8] But much of this has to do with historical accident and with the physical site of the downtown area. The shiny new corporate plazas are found mainly on Bunker Hill, which looms high above the traditional (and now largely Latino) downtown. In spite of the reopening of the Angels Flight funicular connecting the two zones, they are really quite separate. The white-collar, plaza-using people on the hill rarely have to contend with the denizens of the large and growing skid row area on the other side of downtown. Is this because of surveillance and control or are they just too far apart? Are they far apart because of deliberate design or are they located where they are as a result of long-standing, "natural" trends in the sorting out of downtown land uses? Just how much evil intent has there been? Although Los Angeles appears to be an extreme case, there are examples in many American downtowns of corporate plazas isolated from the poorer areas. But that is, in part, because financial districts have become so large that it is only to be expected that they are isolated from other downtown subdistricts.

But even in Los Angeles, the issue can be fuzzy. Other than the tiny Pershing Square and (earlier) Olvera Plaza, the public space of downtown Los Angeles has always been on the street. That is still the case today, and the Broadway shopping district is often so packed with people that it is difficult to walk there. Most of the shoppers tend to be "majority" (so-called minorities are a majority in Los Angeles County) with few of the types of people you might expect to find on Rodeo Drive in Beverly Hills, but does this mean that the streets have become peripheral and marginalized? Business is booming. Has the use of public space declined?

A related issue is "who has a right to be in public (or semipublic) space? During earlier, and supposedly superior, golden ages of downtown public space in times past, it was common for vagrants, drunks, prostitutes, and the generally unruly to be carted off to jail or maybe run out of town. It is either utopian or naive to think that such people should now be welcomed in corporate plazas. Downtown projects often involve investments of hundreds of millions of dollars, and a good address can mean the difference between success and failure. If nearby parks and plazas are known as good places (such as in Rockefeller Center), the buildings gain a certain prestige. If the plaza becomes known for drug-dealing and homeless camps, the project is doomed. Quite apart from any social conscience that designers or city planners might or might not have, the

lenders alone would do all that they could to avoid such a liability. Should lenders have such an important role in urban design? Maybe not, but they do, and so it is unlikely that corporate plazas will ever be known as bastions of equity and diversity. Given these realities, it sometimes seems that diversity prevails even in privatized space. I have systematically observed a great many shopping atria, festival markets, and corporate plazas and have generally concluded that the poor, non-spending public is welcome. Only the very loud, rude, drunk, sleeping-bagged, and unimaginably disheveled are subject to ejection, and this is a tiny percentage of the typical downtown population.

Other Spaces, Other Cities

Even if we wholeheartedly accept the idea that many of the best gathering spots in American downtowns are really privatized and controlled spaces, it does not necessarily follow that "real" public spaces have become marginal and degraded. There are literally thousands of examples of good public space in the hearts of American downtowns, from the state house lawns in Columbus and Denver to Pioneer Courthouse Square in Portland and Pioneer Square in Seattle. Cities such as Pittsburgh and Cleveland have vast green areas in their city centers that did not exist at all fifty years ago. Seattle has created small parks overlooking its waterfront as well as a kind of plaza major in the heart of its retail district. Providence and Indianapolis now have public promenades along downtown waterways for the first time ever. Boston Common is still full of people nearly every day, and so is Washington Square in New York City.

Sometimes problems arise when a critical mass leads to a place-based subculture of "homelessness." If there are so many people on "skid row," as there are on Fourth Street in downtown L.A., that nearby areas feel threatened by possible mass invasions, then segregationist tendencies become more blatant and certain parks or plazas are "set aside" for makeshift shelters. Sometimes the invasions do occur and can be disruptive. When the San Francisco City Hall and its associated civic center were being refurbished in the early 1990s, several buildings were closed and so many homeless men gathered on the plaza each day and night that it became a place to avoid for most city residents. The fact that such numbers exist would seem to be a national problem that goes beyond the abilities and resources of individual city planners and architects to remedy. Until such people are either absorbed into the economy or properly housed and cared for, modifying the designs of downtown plazas will not make the world a much better place.

Civic Architecture and Pride of Place

While the issue of public space has garnered most of the attention from scholars, architecture can also play an important role in creating civic pride and gluing communities together. In the Eurocentric image of the ideal city, city halls, cathedrals, bell towers, city gates, and public squares dominated by monumental civic structures loom large. In American cities, the effort was more diffuse, but still, city halls, courthouses, libraries, and public hospitals were once built to be civic monuments as well as (or some would argue instead of) useful buildings. In Buffalo and Los Angeles, city halls were built in the form of exotic skyscrapers that dominated the skyline physically as well as symbolically. Much of this changed in the decades immediately following World War II. As modernism took hold, city halls and county courthouses increasingly took the shape of nondescript boxes that were practically invisible. In San Diego, for example, the city hall (1964) was actually a fairly expensive structure but it was said to be designed to look inexpensive in order to appease taxpayers. In Seattle, the 1950s-style city administration building is a metallic-blue box that does little to engender a civic sense of place. In Columbus, and many other cities, ancient city buildings remained, but overflow office activities were farmed out to structures scattered around downtown. In addition, as suburbanization flourished during the postwar years, there was little investment in downtown libraries, concert halls, or museums. Municipal buildings often looked tired and dated by the 1970s.

In recent years, much has changed for the better. Portland has a monumental new government building designed by Michael Graves, complete with a hovering statue of Portlandia. Denver, Seattle, Cleveland, San Antonio, Columbus, Charlotte, Indianapolis, and several other cities have new libraries, museums, or performing arts centers downtown, often in association with new or expanded civic open space. Of course, there is the question of just what is a civic building, as opposed to a commercial attraction. Baltimore and Denver have built aquariums downtown while Indianapolis has a new zoo. Many cities also have new "semipublic" stadiums and sports arenas. Most of these will be discussed later under major attractions, but they do have at least some civic characteristics. As yet, few cities have built new and monumental city halls or civic centers, but that could easily be the next phase. However, the example of Boston's brooding, brutalist city hall (circa 1968) set in a cold, wind-swept plaza may be enough to dissuade other cities from following suit.

And now for the rankings:

Variable 3. Civic Space—Grand Buildings, Monuments, Promenades, and Plazas

Indianapolis: A (10)

Indianapolis is a city rich in civic monumentality—downtown is embellished with a state capitol, Monument Circle, a war memorial, art and history museums, green malls, parks, a canal walk, the zoo, a large university campus, and a public market. It may even have too much public space, in that there is no consensus location for major public gatherings (but this would certainly be an unusual problem). Civic architecture competes well with the private scene. The capitol, sports arenas, war memorials, and statuary and other embellishments enhance the sense of the public city. Flags wave, trees grow, and downtown civic identity flourishes.

St. Louis: A (10)

St. Louis has lots of impressive civic monuments, including those dating from many different eras—the courthouse, old city hall, the Gateway Arch, grassy malls, the riverfront park, a library, etc. Sometimes, however, poor con-

Monumental civic space in downtown Indianapolis

Civic mall and courthouse in downtown St. Louis

nections between public spaces, coupled with their large size and lack of enclosure, make them seem too passive and a bit too empty to contribute to a vibrant downtown setting. Still, there are many grand vistas and lots of public
spaces. Some lively in-fill around the green open spaces would help provide urban foci, but already such spaces act as major sites for local recreation and
tourist destinations.

Columbus: A (10)

Columbus is perhaps the best example of a state capitol building and grounds
being the primary focus for a major downtown. Capitol Square is located in the
center of downtown and it is surrounded by major office buildings, theaters,
and shopping destinations. It acts as a sort of "plaza major" for the city. Government office buildings and parks along the river also help to give a sense of
civic spirit and are used as settings for seasonal festivals. City hall is small but
occupies a highly visible setting along the river. The main library and art museum occupy pleasant settings in the downtown core. Small parks act as sub-
district foci for residential projects. The county complex provides a separate
plaza, as do three downtown educational institutions.

The state capitol in Columbus anchors the downtown core.

Denver: A (10)

Denver has a truly monumental civic center consisting of the state capitol, county courthouse, library, and art and history museums all facing a large and pleasant green park/plaza. It is ideally designed to reinforce civic identity but, due in part to a mismatched grid, it seems slightly off-center in relation to the downtown core. The awkward transition is, however, balanced against the view of the capitol accentuated by the grid merge. At the downtown edge (behind the convention center) there is a large university complex (Auraria campus), which includes three public colleges. Street trees, benches, and other embellishments along the Sixteenth Street bus mall also provide a strong sense of the civic.

Cleveland: A− (9)

Cleveland has one of the best examples of a 1930s civic mall with good public buildings and a new library, but it is so big (especially in winter) that it can seem a bit "passive." Public Square is a major focal point although it is too small for everyday use (but good for festivals when closed off to traffic). Linkages be-

Civic Center Park in downtown Denver

tween the downtown core and the waterfront were historically poor but are improving with parks, walkways, light rail, and museums (including the Rock and Roll Hall of Fame) there. New green trails along the railroad tracks in the Flats (the industrial area along the Cuyahoga River) have helped as well. The Cleveland State University campus is on the edge of downtown but could be better linked.

Portland: A− (9)

Downtown Portland has good, inviting public spaces along the Willamette riverfront and the park blocks near the Portland State University campus. Cozy public park blocks provide foci for government buildings in the core such as the city hall and the courthouse. The "postmodern" statues on the Portland Building are intriguing if controversial. There are excellent "human-scale" civic features such as the Ira Keller Fountain and various statuary along the streets. Street trees, bus shelters, murals, and sculpture help to contribute to a sense of civic pride. Pioneer Courthouse Square, a former parking lot, now serves as a very effective plaza major in the center of the retail district.

The Tom McCall Waterfront Park in downtown Portland

Baltimore: B+ (8)

It is hard to classify the Baltimore waterfront since it functions as an excellent public open space, but it must also be counted as a major commercial and cultural attraction. Parks and green spaces nearly surround the Inner Harbor with promenades between the various nodes of activity. There is an impressive city hall facing a somewhat bleak and underutilized plaza. The Washington Monument at Mount Vernon Square anchors the northern end of downtown, but it is increasingly off-center as the life of the city has focused on the waterfront. Public spaces resulting from early urban renewal projects, such as those at Charles Center, seem a bit sterile and dated even with a university campus. Still, there is lots of potential for civic identity and lots of space for strolling.

Providence: B+ (8)

Downtown Providence is dominated by a monumental state capitol high on a nearby hill. Not long ago it was only tenuously linked to the CBD since railroad tracks and open space separated the two nodes of activity. Recently, a shopping mall has been built to link the civic and commercial city and to en-

close the once-vast open area. There is a civic plaza focused on city hall but it is heavily used as a transit center. The county building across the river also adds monumentality. The rebuilt downtown riverfront now serves as a park and promenade. Brown University and Rhode Island School of Design are located close enough to make a visual contribution to the CBD but are not really downtown.

Pittsburgh: B (7)

The park and monumental fountain at the Point (Point State Park at the Golden Triangle) gives the city a readily identifiable icon as well as much needed green space, but it is somewhat peripheral to downtown core activity. There are good civic buildings but they are usually embedded in the urban fabric and do not contribute much to either the skyline or to civic open space. Market Square acts as a sort of plaza major for the office core, and it is reinforced by the plaza at the PPG Building. Duquesne University and Robert Morris College as well as a variety of professional schools are in or near downtown but include little public space. The primary art museum and university district is located in Oaklawn several miles to the east. Riverfront parks are under construction, but except at the Point, river access is still surprisingly minimal.

Market Square is a small civic space in downtown Pittsburgh.

San Antonio: B (7)

Like Baltimore's Harborplace, the Riverwalk is hard to classify since it functions as both a green (albeit small) civic space and as a major commercial attraction. The combination of Riverwalk, Alamo Square, Hemisfair Plaza (with the Institute of Texan Cultures), and traditional Latino plazas, especially at the Mercado, makes for strong city identity and lots of good open space, although the parts are not always well linked or very distinctive or grand. The new library and the old courthouse are pleasant civic structures as well but are off-center, as is the new downtown campus of UT San Antonio.

Atlanta: B (7)

There is a monumental state capitol with associated city and federal office buildings, but these are located so as to be off-center and peripheral to other downtown centers. Woodruff Park is located in the geographic center of downtown but generally fails as a plaza major. The semipublic plaza at the entry to Underground Atlanta provides some district-level civic space at least when the shops are open. Georgia State University is embedded in the older section of downtown but provides little in the way of civic open space. The new Centennial Olympic Park provides a green focus for the CNN Center, the Philips Arena, and the convention center and will no doubt become more important as downtown moves toward it. The potential is there for a grand central public square. Most museums, art galleries, gardens, and parks, however, are in Midtown, and so the CBD core seems overly focused on the business towers and hotels strung out along Peachtree Street. Many of the buildings exude a massive and fortresslike ambience at street level.

Seattle: B− (6)

There are no grand government buildings or civic plazas in downtown Seattle except for the new "piazza" at Westlake Park, which was built as a focus for new retail projects. City hall is dated and dreary. Freeway Park, built over Interstate 5, adds a bit of peripheral greenery and helps to link the downtown core to nearby Capitol Hill. Some small parks have been constructed near Pike Place Market as waterfront overlooks, and the market itself serves as both an attraction and as "public" space. The waterfront is largely a commercial attraction but there are pleasant promenades. A new art museum and symphony hall have recently been completed and add a sense of civic culture to the core. The Space Needle at the

The new Centennial Olympic Park is a potential civic anchor for downtown Atlanta.

1962 World's Fairgrounds (Seattle Center) is peripheral but still serves as a civic icon. Pioneer Square and the associated pedestrian mall act as magnets for preservation activities. Still, there is relatively little civic monumentality, as most of the good spaces are "vest pocket" plazas rather than grand civic spaces.

Minneapolis: B− (6)

There is a grand city hall dating from the nineteenth century but it is embedded in the urban fabric with no civic space around it. Loring Park anchors the southern end of the downtown and acts as a focus for residential uses as well as an art museum, but it is remote from the CBD core. The falls and locks along the Mississippi, which gave rise to the original city, have benefited from some historic markers and riverside trails on the northern side of downtown, but the locks themselves, although interesting, limit the "waterfront park" idea. The bus mall provides a consensus focus but skyways diminish views of possible civic icons. The University of Minnesota is not far away but is poorly linked to the downtown.

San Diego: C+ (5)

The only attempt at a grand civic space that has worked at all is the Art Deco county (originally city and county) building built along a newly dredged and

defined waterfront during the 1930s. It has always been peripheral to the core of the CBD, however. Horton Plaza (the original park) is now sanitized and depopulated and dominated by Planet Hollywood and a shopping mall. Nice parks have been built at the marina but they are walled off from the downtown by the convention center. There are nice street embellishments along Fifth Avenue, but there is also much commercial encroachment. The civic center, which includes a boxy city hall and theater, is a sterile, superblock barrier, although it occasionally works well for special events. There is a community college but it is peripheral, and the main library is small and dated. Museums and theaters are in Balboa Park, which is not far from downtown but far enough to be a separate district, especially since it is on the other side of a freeway. If Balboa Park could be included as "downtown," San Diego's score would soar, but I have got to draw the line someplace.

Phoenix: C+ (5)

The state capitol is located far to the west of the financial core and makes almost no contribution to the downtown sense of place. Patriot Park, reinforced by the old Art Deco city hall, provides some civic space, but boxy government buildings dominate the scene. The new library and art museum are also located well beyond the core in association with the Phoenix version of a grassy park built over a freeway (Margaret T. Hance Park). Symphony Hall and a museum complex (history and science) add civic life to the downtown but are best discussed under major attractions. Both civic activities and icons are generally too spread out and poorly grouped to give downtown a high rating.

Charlotte: C (4)

Charlotte has done a good job of dressing up its downtown streets with art and sculpture but, other than a pleasant library and a green "civic" park, there is little in the way of symbolic public architecture or open space. Even the park acts as a dead zone between major hotels as much as an important civic node. There are some vest-pocket parks that have provided foci for housing efforts and some nice semipublic corporate plazas, but there is little civic grandeur.

Summary: Civic Space

> Indianapolis 10
> St. Louis 10
> Columbus 10
> Denver 10

Cleveland 9
Portland 9
Baltimore 8
Providence 8
Pittsburgh 7
San Antonio 7
Atlanta 7
Seattle 6
Minneapolis 6
San Diego 5
Phoenix 5
Charlotte 4

Conclusion

Downtowns come in a variety of shapes and sizes. The purpose of this chapter has been to explore how physical site characteristics and various aspects of street morphology and civic space can play important roles in delimiting downtown boundaries and shaping downtown patterns. One of the main reasons that, even after decades of homogenizing trends, most downtowns still retain an identifiable character and sense of place is the lasting influence of their physical sites and initial adjustments to them. While urban renewal and highway construction have greatly impacted downtown form, much of the original "footprint" remains. While controversies abound concerning everything from the role and nature of the street to what it means to be "public," these are the city characteristics that provide the stage set for the economic activities that have remained in the downtown core. As I examine individual case studies later in the book, many of the lasting variations in downtown form will become more evident.

Traditional Downtown Functions

Offices, Retailing, Hotels, and Convention Centers

Office Buildings and the Downtown Skyline

In the preceding chapter I described the downtown stage set. Now it is time to bring on the players. For most of the past century, office buildings have been the "big gorilla" in shaping the structure of the typical American downtown. Since office towers can be built to incredible heights, unlike retail establishments or factories that must rely on greater amounts of horizontal space, office space can generate the most profit per square foot of land. In addition, by the early decades of the twentieth century, skyscraping office towers were becoming monumental corporate symbols—giant logos for the huge banks, insurance companies, and industrial firms that were coming into being at the time. Tall office buildings and the resulting skyline came to epitomize what an important city should look like in America. Competition between corporations for the tallest headquarters and competition between cities for the most impressive skylines were under way even before World War I. The skyscraper came to symbolize the energy, enthusiasm, and optimism that characterized the American urban economy in the early 1900s. The fact that this happened in

America rather than elsewhere can be summarized with four topics—cultural values, technological convergence, economic organization, and governmental policies.

While we usually like to pretend that everything is based on the bottom line of economics, cultural values have certainly helped to shape the American urban skyline. When pride is at stake, cities usually build something tall and monumental in order to gain identity and recognition. During medieval times, European cities competed to build the tallest cathedral, and the competition continued into the 1800s as remarkable city hall towers went up from Manchester to Munich. In most European cities, as in many other parts of the world, a satisfactory degree of monumentality had been achieved by the dawning of the twentieth century, but this was not typically the case in the newer cities of America. Lacking cathedrals, castles, city gates, and palaces, many American corporations sought to dress up their hometowns with Gothic and Renaissance towers housing office space. Indeed, the 792-foot Woolworth Building was known as the "Cathedral of Commerce" when it appeared on the New York City skyline in 1913. By the 1930s, skyscrapers, such as the Empire State Building, were among the most famous icons in America, and a skyline symbolized a prosperous city.

Technology also played an important role in the origin and rapid diffusion of the skyscraper office tower. Although the skills needed to build office towers could be found in many countries, it was in America that a wide variety of related technologies came together quickly in a relatively flexible, innovative context. This was especially true where disasters, such as the Chicago Fire of 1871, required the rebuilding of huge areas with larger, more substantial structures. It was in Chicago that the idea of steel frame, or birdcage, construction was developed, allowing buildings of great height to be put up without massive, load-bearing stone walls on the lower floors. As architects and engineers gathered in Chicago, and later in San Francisco, Baltimore, and other fire-damaged downtowns, several new technologies came together. Electric elevators and lights, central heating and cooling systems, telephones, and a variety of business machines facilitated the development of radically new types of office buildings. In addition, the Bessemer process made large quantities of relatively inexpensive steel available. In European cities, and even in some older American ones such as Boston, established ways of doing things and traditional landscape preferences were harder to change.

Cultural values and technical abilities made skyscraper office towers possi-

ble, but such buildings were also increasingly necessary in the new economy of big business. At the turn of the twentieth century, the giant corporation came into its own with large numbers of office workers and records headquartered in one place. By the 1920s, the number of office workers had exploded, with hundreds if not thousands of people working away on typewriters, adding machines, and other office innovations all in one building. In Europe and Japan, corporations were far more likely to be associated with well-known families and to serve smaller, politically fragmented regions where they were part of the accepted establishment. In America, newly formed giant corporations serving a newly rail-accessible national market were more common, and many gained not only needed office space but name recognition through the construction of famous towers.

Government policies also helped to make the American skyscraper possible. Tall buildings have always created controversy, and many European and Asian cities established strict height limits as early as the seventeenth century in order to manage congestion and ensure the availability of adequate light and air at street level. From Paris to Vienna, uniform profiles of six-to-eight-story buildings dominate the urban cores to this day. A few American cities such as Boston and Washington also adopted strict height limits, at least for a while, but most chose to facilitate or even to encourage skyscraper construction.

In order to avoid overloading the infrastructure and to control the densities that rampant skyscraper competition could bring, by 1916 the first zoning laws were established that related the allowable height and bulk of buildings to the size of their lots. These "floor area ratios" or FARs varied from district to district. In the core of the CBD, for example, a FAR of fifteen would mean that a fifteen-story building could be constructed over the entire lot or a thirty-story building over half the lot or any combination that would lead to the desired 15:1 ratio.[1] Outside the downtown, much lower FARs were the norm. As a result, skyscrapers came to dominate a small financial core that was surrounded by smaller buildings. Zoning thus helped to accentuate a steep land-value gradient from a high-density core to an underutilized periphery. Zoning laws varied from city to city, but most governments created at least one district that permitted a "skyline" of high-rise buildings.

Over the decades, government policies have continued to facilitate the adoption of the skyscraper. The urban renewal projects of the 1950s and 1960s, for example, nearly always emphasized the construction of gleaming new office towers that would demonstrate to the world that the city had "turned around."

Typically, governments offered subsidized land, lot assembly by eminent domain, and tax abatements in order to make downtown land appealing for office construction.

During the past century, the four contexts—cultural, technological, economic, and governmental—have combined to embed the skyscraper office tower and the urban skyline in the personality of the American downtown. While more recently monumental hotels and residential towers have served to embellish the skylines of many cities, the office tower remains the primary symbol of successful downtown revitalization.

Office Buildings and Downtown Viability

Office buildings became important shapers of downtown morphology only during the past hundred years. There was simply very little office employment in most cities until the late 1800s, and even then office space normally existed in the shadow of more traditional activities such as retail, wholesale, warehousing, and even manufacturing. Offices over shops or in a corner of a warehouse were typical, and many important meetings took place in coffeehouses during the early 1800s. Things had changed by the 1890s, and office towers were beginning to appear that exceeded twenty stories. In 1915, the Equitable Building, a thirty-nine-story tower in lower Manhattan, became the first office building to have over one million square feet of office space. For the first time, specialized financial districts emerged in larger cities to occupy the most expensive and centrally located land. Residences, factories, and even department stores were relegated to more peripheral sites. Famous projects such as Rockefeller Center in New York City, the Terminal Tower in Cleveland, the Russ Building in San Francisco, and the Tribune Tower in Chicago came to epitomize downtown urbanity.

During the first half of the twentieth century, however, office districts continued to share downtown space with department stores, port facilities, theaters, hotels, food markets, and even remnant residential areas, but by 1950 most of these other activities had diminished in importance. As suburban shopping malls, motel strips, apartment districts, industrial parks, and drive-in restaurants and movie theaters proliferated, downtowns began to concentrate on the one thing they still seemed to do well—office buildings. Beginning with the urban renewal projects of the 1950s and through the building boom of the 1960s, most American downtowns lost employment in nearly every category except office functions. City halls and county courthouses expanded with

The Terminal Tower Complex (1928) in Cleveland has office space, department stores, hotels, and a commuter rail line.

increased government employment. They also continued to attract law firms and other related office activities to nearby locations. Many downtown banks and corporate headquarters had so much already invested in downtown land and buildings that they carried out large construction projects in order to protect their investments.

Most of these efforts were conservative at first and few monumental towers were constructed during the 1950s, but by the late 1960s the race for the sky was back. The John Hancock Building in Chicago (1969) was the first to soar above one thousand feet since the completion of the Empire State Building in 1933. Soon plans were announced for the twin World Trade Center Towers (1973–2001) in New York and the Transamerica Tower in San Francisco. Office towers were becoming bigger and more expensive than ever, with buildings of over a

Fortresslike blank walls protect the Westin Hotel in Atlanta.

million square feet going up in major cities around the country by the early 1970s. While these towers had tremendous symbolic power, they often served to implode downtown activity into just a few core blocks. With in-house cafeterias and connected or underground parking garages, the new office towers were sometimes in downtown but not part of it. Employees could drive in at dawn and out at sunset but never walk on a city street. By the 1970s, many downtowns were perceived to be dangerous places, and so blank, fortresslike walls were often constructed at street level to "protect" workers from the evil city.

Still, in spite of these design problems, the new office towers represented a commitment to and a huge investment in the downtown by those with money and power. More importantly, they brought employment to the central city that kept alive at least some struggling restaurants and shops. A good downtown almost had to have office towers. Even people who seldom went downtown could see the skyline and feel that the city was still alive.

The Postmodern Boom of the 1980s

The towers that resulted from the boom of the 1980s tended to be more sensitive to urban design problems. In general, downtown buildings became more

interesting and engaging from top to bottom. At the top, the simple glass box was increasingly embellished with turrets, spires, and towers inspired by the famous Art Deco buildings of the prewar decades. Individual buildings and city skylines became more recognizable as skyscrapers added a sense of architectural whimsy to "form follows function" modernism. Many of the new buildings were illuminated at night in order to show off architectural details. In cities as diverse as Cleveland, Charlotte, San Diego, St. Louis, and Philadelphia, the tallest skyscraper in the city resulted from the postmodern boom of the 1980s.

The postmodern towers were often more interesting at street level as well. Many cities encouraged food courts and retail establishments, often in association with (semi)public plazas or internal atria. The construction of a traditional urban "street wall" of doors and windows was also encouraged over the modernist approach of one door set back from the street. Sometimes theaters, art galleries, and pubs were included in office towers, as mixed-use buildings became popular once again. In a few cases, residential, hotel, and retail activities were all included in one building. Decorative doors, statues, and lighting helped to make office lobbies increasingly interesting as well. Of course, some of these design elements have caused controversy in that the well-designed "public" spaces at the bases of the new towers are really private and exclude part

A private/public plaza in downtown Charlotte

of the downtown population. Still, postmodern design elements have helped to dress up and enliven the once sterile financial districts of many American downtowns.

The Boom Ends: Rethinking Skylines in the 1990s

By the early 1990s, the latest boom in office tower construction was over. Many downtowns had a glut of office space, especially those in regions experiencing downturns in the local economy. In addition, the maturing of the suburban "edge city" office park concept into a more viable option for major corporate headquarters caused some major office employers to forsake downtown locations. The choice was no longer simply between a downtown setting and a remote suburban hideaway since many edge cities offered amenities and opportunities once found only in traditional downtowns. In still other cases, local resistance to unbridled skyscraper construction led to office caps, lower FAR guidelines, or other obstacles to further growth. Authorities in cities such as San Francisco and Seattle began to warn against "Manhattanization," meaning the oppressive scale and functional uniformity associated with too many office towers concentrated in a small financial district. The World Trade Center attack in 2001 may further lessen the desire for landmark structures.

By the late 1980s, it was becoming abundantly clear that good downtowns could and should have a lot more going for them than plentiful office space. For nearly three decades, downtown planners and entrepreneurs had put all their eggs in one basket, emphasizing office projects at the expense of everything else. However, the 1990s focused attention on many new and different downtown projects, from sports arenas and upscale housing to waterfront parks and cultural centers. Office buildings sometimes became something to de-emphasize rather than promote. For a variety of reasons, very few major office towers went up during the 1990s. Perhaps ironically, one of the tallest American "skyscraper" projects to be completed between 1995 and 2000 was the New York skyline in Las Vegas.

Preservation and renovation have also impacted the downtown core. During the halcyon days of urban renewal, older buildings were usually viewed as obsolete eyesores and torn down whenever possible. In recent years, a preservation ethic has emerged that has led to the remodeling of many older downtown office buildings and often their conversion to other uses. Even in Lower Manhattan, America's original "downtown," many older towers that are considered too difficult to upgrade and rewire as competitive office space have been

converted to apartments and condominiums. This, it would seem, is a logical next step. Many buildings once used for warehousing and industry were successfully converted into loft apartments when the demand for that space turned soft decades ago.

Educational, civic, and recreational uses have found space in older office structures as well. In some cities, universities, museums, theaters, and retirement centers have filled space once used by banks and insurance companies. These older buildings add architectural variety and detail as well as a sense of maturity to the financial core.

What Makes a Good Downtown Financial District?

Like it or not, in order to be successful American downtowns still need to have lots of office space. Office workers provide a captive market of sorts for a variety of downtown support activities such as restaurants, specialty shops, and clothing stores. The presence of upper-level management in downtown offices usually means a commitment by the power elite to city plans and programs aimed at improving the quality of the infrastructure and the staging of special events. The question in middle-sized downtowns is how many office buildings and office employees are necessary for a viable financial district. A good guess might be that a goal of around twenty million square feet of office space and perhaps 150,000 office employees is something to shoot for. Downtowns with a lot less space tend to have too few buildings, resulting in large gaps in the urban fabric. Downtowns with a lot more tend to have sterile financial canyons and serious problems with rush hour commuting, unless they have good public transit systems.

Not only do good downtowns need a significant amount of office space, they also need a reasonable number of good, highly visible buildings—in short, skylines that can give sprawling metropolitan areas a strong sense of center. Ideally, a good downtown is made up of office buildings in a variety of styles, sizes, and ages dating from the 1890s to the 1990s. If all the major buildings are either too old or too new, the downtown may lack the strong sense of place that comes from the unique assemblage of elements from different architectural and planning eras. The best downtowns have major structures that serve as icons for the various ideologies and economic considerations that have shaped the center city over time.

A good skyline is not only one that is made up of good buildings; it also has a certain pleasant and impressive composition. The whole may be greater or

less than the sum of its parts if the buildings are nicely grouped. While some skylines are impressive simply because of their mass, too many glass boxes of similar height can result in a skyline "wall" with few interesting peaks or valleys. On the other hand, a much smaller group of buildings can serve to create a memorable skyline if a few remarkable towers stand out against a backdrop of supporting but pleasant smaller structures. Of course, it also helps if there are good places from which to view a skyline—as from across a river or from a nearby hill. In this sense, cities like Pittsburgh and Seattle have an advantage over cities with fewer view sites.

With all of this in mind, we can begin to evaluate our sample of American downtowns with regard to their total office space, architecturally interesting buildings, and overall skylines.

Variable 4. Skyline Image and Amount, Variety, and Quality of Office Space

Cleveland: A (10)

Cleveland has a spectacular skyline made up of monumental and easily recognizable buildings from several eras—especially the Terminal Tower (1928) and the Key Bank (1991), two of the tallest towers in the country. There is a good clustering of towers around Public Square that tends to bring together civic and financial district identity. Downtown is a strong center of office employment in both the public and private sector, with a total of about twenty-two million square feet of office space. Some modernist boxes in the Erie View area add little to the skyline, but at street level the empty 1960s plazas are being filled in with galleries. There is also a good supply of older, historic buildings for conversion and remodeling, which adds to the urban character. Some older buildings have been saved and renovated by using transferred development rights. There are good views of the skyline from the lakefront and from nearby neighborhoods. A basement-level light rail station reinforces the importance of the Terminal Tower Complex on Public Square. Major headquarters include the Key Bank, BP America, and Standard Oil. There are twelve towers over four hundred feet tall, with the tallest at an impressive 950 feet.

Seattle: A (10)

Downtown Seattle has a stunning skyline enhanced by its hill location above the harbor. There are twenty-three buildings over four hundred feet tall, with

the tallest at 943 feet (but actually over a thousand feet above the waterfront). There is a nice mix of interesting postmodern towers and modernist boxes but a relative dearth of old towers except for the five-hundred-foot (1914) L. C. Smith Building. The financial district is dominated by bank headquarters and has over twenty-four million square feet of office space. The towers are nicely clustered in a clearly defined financial district, although steep hills sometimes make intradistrict walking awkward. There are some nice plazas and interior spaces with retail and food venues. Observation decks are accessible on two of the tallest buildings. Skyscrapers have been controversial in Seattle, and there has been public resistance to unlimited tower construction. Numerous hotel and residential towers enhance the city skyline, especially at the northern end.

Pittsburgh: A (10)

Some of the earliest examples of "skyscrapers" (twenty stories) outside of New York City were built in Pittsburgh in the first decades of the twentieth century. There are also several interesting towers from the 1920s. Modernist boxes from the urban renewal period of the 1950s and 1960s and postmodern corporate headquarters from the 1980s complete the skyline. There are seventeen

The Seattle skyline

Corporate towers in downtown Pittsburgh

towers over four hundred feet tall, topped by the 841-foot U.S. Steel (now USX) Building. Other headquarters buildings include Gulf, Pittsburgh Paint and Glass, Mellon Bank, and Alcoa. Interesting towers combined with good clustering and stunning skyline views from the rivers and nearby hills give identity and personality to the downtown. With over eighteen million square feet of office space, downtown Pittsburgh remains a strong employment center even in a relatively slow-growing region. Many buildings provide nice street-level amenities, and there is a good supply of old buildings for restoration.

Atlanta: A (10)

Downtown Atlanta has some excellent postmodern towers, dominated by the NationsBank tower at 1,023 feet. It is difficult to say how many tall buildings are located in downtown since most are on the Peachtree corridor that extends into Midtown and beyond. Still, there are at least twenty towers over four hundred feet and well over thirty-two million square feet of office space in the greater downtown Atlanta region, with seven buildings over seven hundred feet tall. The skyline is less impressive than the number of buildings suggests, however, since the towers are strung out over a great distance and because there are

Modern and postmodern towers in downtown
Atlanta

few good places for a superb skyline view. Banks and headquarters such as
Georgia Pacific and Bell South Telephone dominate the financial district, but
hotels contribute to the skyline as well. There are no significant old towers
but several new ones have deco-like personality. Many of the buildings are
fortresslike with few street-level amenities, although there are some monumental lobbies and interior spaces. Some older buildings are being converted
to residential and university uses. Still, downtown (with Midtown) Atlanta remains a major office center for the Southeast.

Minneapolis: A− (9)

Downtown Minneapolis is dominated by three impressive towers of 775 feet,
with a total of twelve buildings of over four hundred feet. There are also some
very nice older buildings, such as the Foshay Tower (1929). The skyline is made

A compact skyline and the Humphreydome in Minneapolis

up of a mix of modernist and postmodern "retro" towers, and they are nicely grouped in a compact financial district thanks, in part, to the perceived need to be on the skywalk system of second-level walkways. There are nice views of the skyline from the river to the north and east and lakes to the south and west. Skyways connect the office towers to shopping centers and parking garages but also interrupt views of downtown buildings from the street. There are many nice interior spaces (such as the IDS courtyard) which function as urban meeting places. The downtown contains over eighteen million square feet of office space and is the headquarters for Pillsbury, Multifoods, and several banks.

Columbus: B+ (8)

Downtown Columbus contains a good mix of older, modern, and postmodern towers in its financial core at Capitol Square. Various architectural eras are represented by the "deco" Leveque Tower (1927), the modernist Rhodes Office Tower (1973), and later postmodern structures. There are ten buildings over four hundred feet tall, with the tallest at 624 feet. There is a total of about fourteen million square feet of office space. While Capitol Square provides a strong sense of center, there are competing office nodes at the northern and southern ends of downtown, and this dilutes the overall impact of the towers

and also makes for gaps in activity at street level. Several of the towers are well lit at night and include lobbies with theaters, art galleries, and green atria. There are nice views from across the river. Headquarters include Nationwide Insurance, Borden, and American Electric Power, along with banks and state offices.

Denver: B+ (8)

Downtown Denver contains over twenty-two million square feet of office space and fifteen towers of at least four hundred feet. The tallest building is Republic Plaza at 714 feet, but two others are nearly as tall. While the skyline is very impressive in some ways, it is disappointing in others. Most of the towers were built at roughly the same time (early 1980s) as modernist boxes with relatively little personality. There are few older or postmodern skyline elements to add variety to the mix. There is a good linear cluster of buildings making up

Modernist towers form a financial canyon in downtown Denver.

the financial district but few excellent vista points. Since livelier activities are concentrated on the Sixteenth Street Mall, the office core seems a bit sterile, with little interest at street level. Banks and oil-related companies such as Amoco dominate.

Charlotte: B (7)

Downtown Charlotte is dominated by one spectacular postmodern tower (NationsBank now merged with Bank of America) at 871 feet. There are a few nice secondary structures, including the impressive First Union Tower, but not quite enough to fill out the skyline yet. There are no older towers or even older mid-sized office buildings to add architectural variety, since the traditional main street has been largely rebuilt. There are seven buildings of more than four hundred feet and about ten million square feet of office space. Skyways, which in Minneapolis help to tie downtown together, seem unnecessarily plentiful here. Recent court decisions may slow the boom, as some residents have objected to hyperdevelopment. There are some very nice atria and artistic touches at street level that serve to brighten the core ambience a bit.

New towers anchor the compact core of downtown Charlotte.

Indianapolis: B (7)

The compact office core of downtown Indianapolis contains five office towers of more than four hundred feet and is dominated by the 720-foot Bank One Center. There are no older towers but there is a good mix of mid-rise structures that gives the financial core a sense of mass. The office towers are clustered around a consensus core at Monument Circle and this adds to financial district identity. Downtown contains about ten million square feet of office space, supported by the massive Eli Lilly Office Park south of the downtown core and on the other side of the tracks. The financial district is well integrated with retail attractions and civic space.

Baltimore: B (7)

Downtown Baltimore's skyline features modest mass and height for an older, major city, with five towers over four hundred feet, the tallest being 529 feet. Still, there is good clustering, and so there is a nice skyline view from the harbor even though Charles Street, the main drag, has historically pulled the CBD away from the waterfront. With about fourteen million square feet of office space, downtown Baltimore clearly plays second fiddle to nearby Washington, D.C., and is increasingly known more for its harbor amenities. There is a reasonably good mix of old and new towers, including the (1929) Maryland National Bank, but few corporate headquarters. A major fire at the turn of the twentieth century and urban renewal during the 1950s and 1960s deprived the downtown core of much of its character compared to other neighborhoods nearby. Recent attempts to brighten the financial core are having some impact, but most of "the action" is heading for the waterfront.

St. Louis: B (7)

Downtown St. Louis contains a wide variety of old and new buildings in a variety of conditions. There are some very good individual old and new towers but they are quite scattered, with the central core dominated by several very old and often semiderelict structures, thus diluting the sense of a financial district. The Gateway Arch, the tallest structure in the city at 630 feet, dominates the skyline. There are five office towers more than four hundred feet tall, with the tallest at 593 feet. There are good views of the downtown from across the river, but there are few good places to linger in East St. Louis. Downtown con-

tains about fourteen million square feet of office space, with large numbers of older buildings available for renovation as offices or loft housing. Most of the office buildings are fairly well integrated with other downtown activities, although financial district identity is sometimes overwhelmed by civic monuments or sports facilities.

San Diego: B− (6)

There is a very nice skyline view from San Diego Bay, but it is dominated by interesting condos and hotels rather than by office towers. Of the six downtown buildings over four hundred feet tall, only two are exclusively devoted to office uses, with two hotels, two residential towers, and one mixed-use tower filling in. Most of the nearly twelve million square feet of office space is housed in modest glass boxes, although the pointed top of One America Plaza (500 feet) stands out as more unusual, along with the chisel-topped Hyatt Regency and the mixed-use Emerald-Shappery Center. Because height is limited by the presence of the nearby international airport, no tower exceeds five hundred feet. The financial district is cut in two by a superblock civic center and (ugly) municipal prison, thus diluting the sense of a strong downtown core. Still, the skyline is very nice at sunset. As in Baltimore, the downtown seems increasingly geared toward recreation as much as business.

Portland: B− (6)

Downtown Portland does not have an impressive skyline, although, when viewed from either the river or the nearby hills, the core area seems large and vibrant. There are only three towers of more than four hundred feet (with the tallest at 546 feet), and only one, the mixed-use Koin Center, has much individual personality. The towers are also poorly grouped, with one far off on the northern edge of the downtown. Although there are no older major towers, there is a large and well-maintained stock of older office buildings that give character and variety to the city at street level as well as providing a good sense of mass and housing for its ten million square feet of office space. The Portland Building by Michael Graves, with its famous statue of Portlandia, certainly qualifies as unusual office space, but trees often largely hide it. There is little apparent interest in the kind of "skyline competition" found in many American cities, and abundant trees and required street-level retail tend to soften the sense of a bustling financial core.

Phoenix: C+ (5)

Like Atlanta, Phoenix has a downtown that is strung out over a spine (Central Avenue) that is several miles long, and so it is relatively difficult to define a classic "downtown core," although the city has tried to do so with its official designation of a smaller core area called Copper Square. In addition, state office space is off to the west in a separate government node. Several boxy modern buildings of medium height dominate the core of a CBD, which includes private and government subdistricts. The majority of "downtown" office space is on the northern section of a spine heading out of the CBD core. There are only two buildings of more than four hundred feet, but there are dozens of mid-rise structures along Central Avenue, bringing the total amount of office space in the greater downtown area to about fifteen million square feet. There are no real signature office towers either new or old. As in many newer, western cities, downtown Phoenix seems more impressive if you are traveling in a car rather than walking.

San Antonio: C (4)

Downtown San Antonio has a skyline with a lot of character but not much office space. The skyline is dominated by the 622-foot Tower of the Americas, from the 1968 World's Fair, and the nearby Marriott Rivercenter Hotel (546 feet). There are some very nice towers from the 1920s scattered around the city center but only one of them (Tower Life, 1929) contains much office space. There are only two office towers over four hundred feet in downtown, with only about three million square feet of office space, and there is very little new office construction. The downtown is well on its way to becoming a place for recreation rather than business. A small office district on the northern edge of downtown seems rather removed from the life of the city.

Providence: C (4)

The downtown Providence skyline looks good from the state capitol building located on a hill just to the north, even though it consists of little more than three nicely grouped buildings—one a hotel. Only two office towers exceed four hundred feet (the tallest being 428 feet), and there is only about two million square feet of office space downtown. Still, the "big three" consist of a nice mix—an older tower from 1927, a modernist box, and a postmodern hotel. Many interesting older office buildings that were once part of a financial core

San Antonio has relatively few office buildings but lots of tourist attractions.

have been converted to educational institutions or art galleries. Providence has become a satellite of Boston and it may play some role in the future as a sort of edge city. At street level, the downtown core has a certain charm, especially where the facades of older buildings have been maintained.

Summary: Skyline and Office Space

> Cleveland 10
> Seattle 10
> Pittsburgh 10
> Atlanta 10
> Minneapolis 9
> Columbus 8
> Denver 8
> Charlotte 7
> Indianapolis 7
> Baltimore 7
> St. Louis 7
> San Diego 6

Portland 6
Phoenix 5
San Antonio 4
Providence 4

Downtown Shopping: Department Stores, Malls, Festival Centers, and Street-Level Shops

Throughout history, the main activity in most urban places was shopping. The market square typically occupied the most central location in European cities and was busy every day. Gradually, permanent shops grew up around the central square to complement the churches and civic buildings that shared the location. American cities had a more difficult time maintaining the market square tradition, and many were either ignored or built over as fickle shoppers migrated to the more prestigious streets. Since shopping and improvements in transportation, such as the horsecar and streetcar, evolved together in American cities, shopping districts were often rather linear. Fifth Avenue in New York epitomized the tendency. As stores got bigger and required larger lots, they usually migrated away from the congested waterfront or CBD core and toward the elite residential areas. Still, well into the twentieth century, high-end retailing was the activity that could pay the most for street-level space, and so most office buildings and hotels sought first-floor stores. Mixed-use buildings were the norm.

In both European and American cities, there were early attempts to disassociate retailing from the often messy and unpredictable street. Borrowing from schemes first tried in London and Paris, several American cities constructed enclosed shopping arcades of one or more stories. The Weybosset Arcade in Providence (1828) is the oldest arcade still in existence, while the Cleveland Arcade (1890) is by far the largest. Arcades typically rented small spaces to individual shop owners and so maintained some of the diversity of the traditional street. With the advent of the arcade, however, people could wear their finest clothes and shop and promenade away from the mud and riffraff of the city.

By the middle of the nineteenth century, the department store evolved to further consolidate major retailing into large, multistory buildings with each floor specializing in different types of goods. Unlike arcades, department stores were under one management, and so goods could be moved from floor to floor depending on how well they were selling. Bargain basements gradually evolved for slow-selling items. By the early 1900s, some of the larger stores surpassed

The Weybosset Arcade (1828) in downtown Providence

one million square feet of space spread over as many as ten floors and included not only every product normally associated with downtown shopping but also a variety of restaurants and cafes. Some even had lecture halls and rooftop amusement parks. Over the decades, electric elevators, escalators, heating, lighting, and cooling systems helped to internalize retail activities into large, self-contained structures. By the 1930s, windows were often covered over in order to gain more space for storage and display. Most downtowns were dominated by two or three huge, locally owned stores, located within a block or two of each other and anchored an increasingly distinctive retail district. Street-level shops occupied the spaces between the large stores but were most often small and highly specialized compared to the department stores. Pharmacies, card shops, shoe stores, and soda fountains were typical.

As stores became larger and more self-contained, peripheral shopping streets became less competitive. Many once vibrant streets came to be known as skid rows as bars, pool halls, used clothing stores, magazine and tobacco outlets, and repair shops predominated. By the 1950s, many of these marginal streets were slated for urban renewal and cities stopped maintaining the infrastructure. Building owners were denied loans for mortgages and expansion. The image of downtown as sleazy and even dangerous emerged, and middle-

class shoppers increasingly avoided certain streets. As parking garages were built to serve the large department stores and fewer people supported the stores near bus stops, street-level retailing declined further.

Even in the so-called golden ages of downtown shopping, most people visited only a very small section of downtown and often spent much of the day in just a few major department stores. The glory of once vibrant streets such as Market Street in San Francisco or Euclid Avenue in Cleveland faded by the 1950s as downtown shoppers focused on the offerings in Macy's or Higbee's department stores. The issue of street life has long been controversial in the typical American downtown.

The Struggle to Keep Downtown Shopping Alive

Retail activities all but died in most American downtowns during the 1960s and 1970s. By the early 1980s, downtowns could be divided into those that still had at least one major department store—such as Kaufmann's in Pittsburgh, Lazarus in Columbus, or Mier and Frank in Portland—and those that did not. In the former, downtown remained at least a minor shopping destination, while in the latter, retailing all but disappeared. Phoenix, San Diego, Provi-

The huge Lazarus department store anchors the retail core in Columbus. It is now connected to a shopping mall.

dence, Denver, and Charlotte all went through a period with no major downtown store. Some are still in this period.

Part of the problem was that most department stores ceased to be locally owned and managed. As huge department store chains such as Federated formed, or as a few successful stores such as Nordstrom expanded into new markets, individual downtowns lost control of their retail fates. Decisions to close or revitalize stores were made elsewhere, on the basis of market research data, and civic loyalty seldom mattered. Most people, it seemed, would never be willing to shop where they had to pay to park. Department stores without national connections and sophisticated strategies were often the first to close, and so the retail district imploded around one or two remaining extralocal stores. Meanwhile, street-level retailing either dried up entirely or focused on a few remaining stalwarts such as Woolworth's and shoe repair shops. In some cases, antique stores, flower shops, and jewelers managed to hang on, especially in and around fashionable hotels and major office buildings, but these seldom constituted anything approaching a retail district.

Retail Returns: The Invention of the Urban Shopping Center

The earliest suburban shopping centers were usually little more than crescent-shaped strips focused on grocery and hardware stores and service establishments. By the 1960s, however, it was common to have large suburban malls with two or three department store anchors supported by a number of specialty shops, restaurants, and even movie theaters. As design ideas evolved, more malls were enclosed and filled with amenities such as fountains and gardens. As long as malls were modest in size, they were normally one story, with anchor stores of possibly two or three floors. Most were located in a sea of free parking in easily accessible suburban locations. As shopping malls got larger, serious design changes were necessary.

Since most American shoppers are loath to walk more than about six hundred feet unless they have to, as malls increased in size, they often went vertical so as to increase visual contact with and elevator access to more stores.[2] Visitors could walk in the door and immediately see a dazzling array of stores on three levels. People seemed more than willing to take escalators and elevators to higher floors in order to seek out the shops they wanted to visit. In addition, as mall size increased, the amount of space needed for parking also increased, quickly exceeding the area people were willing to negotiate on foot. The only answer was to build multilevel parking garages. A final change in mall design

and organization focused on the addition of various support activities at the peripheries of mall sites. Free-standing restaurants, office buildings, and service establishments appeared near the once fortresslike mall structures. The result of all this was the somewhat accidental invention of the "urban" mall. While traditional suburban malls were far too space-extensive to be built downtown, the new multilevel malls with parking garages and skyways to adjacent structures were perfect for downtown retail districts. This was especially true where older department stores and their garages were still functioning or where urban renewal had made large, contiguous lots available.

Still, there were some design problems to work out before urban malls could be attractive to a suburban clientele, often wary of the marginal character of older downtown retail districts. Urban malls had to appear to be both upscale and well connected to other desirable and safe attractions. Often the new malls borrowed from Victorian shopping arcades by having ornate skylights, sometimes with views of nearby office towers. Where possible, they were connected to existing department stores, hotels, and performing arts centers so that people could spend time in the mall during multipurpose downtown trips. Special events and activities such as noontime concerts, reminiscent of the heyday of the department store, were added in order to attract nearby office workers. Food courts were expanded into attractive destinations for downtown workers as well as shoppers. Multiplex cinemas and other entertainment venues were sometimes included in the new malls as well. Where possible, prestigious anchor stores new to the metropolitan area were brought in to increase the novelty and appeal of the downtown shopping experience.

The trick was to design the malls so as to be large, exciting, and urban but still sufficiently enclosed so as to be protected from the still suspect city. Gradually, as the images of many downtowns have improved, the malls have opened up in order to improve connections to nearby street-level shops and cafes. More recently, outward-facing stores have been built into the street walls of the mall structures in order to minimize the urban fortress syndrome. The trend toward whimsical postmodern design and away from the stolid seriousness of modernism has also helped to create and sell the idea of downtown as a fun and attractive place well worth some discretionary time and money. The evolution of the downtown shopping mall continues as cities and developers search for the right combination of design, scale, location, and financing.

As the urban mall concept became more sophisticated, variations on the theme were included in other types of buildings such as office towers, hotels,

convention centers, and even sports arenas. Planning authorities encouraged mixed-use projects through the use of zoning bonuses and other carrots, and some even managed to require street-level retail outlets in the downtown core. The results have been urban minimalls in the lobbies of office towers complete with attractive fountains and upscale coffee bars. Developers have gradually come to see vibrant lobby space and attractive plazas as important attributes of good and profitable new towers. Some downtown office workers avoid rush hour traffic by coming in early to have breakfast (or even to work out in an in-house fitness center) or staying late to have a drink and listen to jazz.

In many cities, such as Columbus, Indianapolis, and San Antonio, new malls have helped to bolster and revitalize older retail districts with their venerable department stores and support facilities. In a few cities, such as Seattle, Portland, and San Francisco, downtown retailing has become so successful that new, freestanding department stores have been built, in most cases for the first time in nearly a century. In other cities, such as Providence or Baltimore, new malls have been constructed in areas that were never part of a retail district but are big enough, either alone or in combination with other attractions, to create one from scratch. In still other cities, such as Cleveland, new urban malls have been built in conjunction with commuter rail systems and sports complexes. Several downtowns, however, have yet to establish a viable retail environment. From the standpoint of the average city resident, this is one of the most important factors in evaluating downtown success.

Festival Markets and the Downtown Sense of Place

Department stores and malls represent the revitalization of traditional downtown retailing, but there is another important dimension of the downtown shopping revival. The festival market idea derives as much from the European medieval market square with its colorful combination of jugglers, clowns, and open-air carts as from the enclosed tradition of arcades and department stores. These were once part of the American market scene as well and still lingered on in a few places such as Quincy Market in Boston at mid–twentieth century. However, they were generally thought to be quite passé in the drive-in world of the 1960s. Perhaps the concept had to wait a few years until most memories of earlier open street markets were forgotten in order for it to be brought back as a romanticized novelty.

With the combination of urban renewal, waterfront revitalization, deindustrialization, and a new appreciation for historic preservation during the Bi-

The Tower City Shopping Mall in Cleveland's Terminal Tower

centennial decade of the 1970s, the way was paved for the rehabilitation and reuse of old warehouses and factories as well as the construction of new sympathetically designed structures. Although a variation on the idea was pioneered in San Antonio as early as the 1920s, the popularity of festival markets really took off during the 1970s. With Ghirardelli Square (1964) in San Francisco and Quincy Market (1976) leading the way, cities all across America sought to create specialized retail districts that emphasized unusual settings, experiences, and fun as much as everyday shopping bargains. The idea was to build on local history and sense of place by emphasizing physical site, old buildings, views of the city, and local historic themes and cuisines. The Rouse Corporation played a major role in diffusing design ideas from its headquarters in Baltimore. Indeed, Harborplace in Baltimore (1980) epitomizes this "fun zone" approach to retailing, and it is now one of the most successful urban tourist destinations in the country.

Most, if not all, of the shopping in these marketplaces is nonessential and depends on fairly large doses of discretionary income and tourism to keep it going. Shops selling kites, candles, and T-shirts may predominate. Some social critics and preservationists have criticized the blatant commodification and hokey inauthenticity common to most of the theme "villages," but there is lit-

The former Union Station in St. Louis is now a vast shopping and tourist complex.

tle doubt that the more successful ones have become extremely popular and profitable. Some have failed, however, and others may suffer as the novelty wears off. To be successful, I would argue, such places need to be not only well designed and managed, but well integrated with other downtown districts.

Festival markets have also flourished in locations other than along waterfronts or other amenities. One of the largest is Union Station in St. Louis. The huge train station is located on the edge of downtown in what was once a largely rail-oriented industrial district. Many other cities including Cincinnati and Chattanooga have also used railroad stations for theme retailing.

Not all old warehouses have been rehabilitated as themed fun zones. In New York City, many former warehouses are now used for urban variations of big-box retailing such as K Mart and Crate and Barrel. The large spaces and loading docks make it relatively easy to insert suburban-scale retail activities into the heart of the once congested city. Of course, in all kinds of districts, old factories and warehouses have also been converted to brewpubs and restaurants.

Traditional Street-Level Retailing: The Toughest Nut to Crack

After decades of enclosing design strategies coupled with rhetoric that basically denigrated the street (street people, street walkers, street urchins, street

smarts, etc.), it is no wonder that successful downtown retailing first returned to relatively large, self-contained settings. Only recently have planners and urbanists begun to seriously decry the death of the street and the social evils associated with segregation by enclosed retail venues. How can we say that downtowns have been revived, they say, when there is no one on the street? Some cities have gone so far as to require street-level retail on certain streets, but there are other factors at work as well. Downtowns that have good mass transit systems including light rail and buses have seen the popularity of specialty shops and coffee outlets increase near stops, especially along bus malls such as those in Denver, Portland, and Minneapolis. The conversion of many former skid rows to specialized art and antique districts has also increased the number of strolling shoppers. Of course, this raises questions about where the former skid row denizens are now able to hang out.

While street-level retailing does seem to be slowly gaining popularity, from First Avenue in Seattle to Sixteenth Street in Denver, there are still relatively few "great streets" in American downtowns, at least as defined by the number of viable retail establishments. Even in larger cities such as New York and Boston, busy shopping streets like Fifth Avenue and Newbury Street are the exception rather than the rule. But then, they probably always were.

Perhaps American shoppers do not like to stray too far from the safety of major nodes of activity. In some downtowns, the best streets for strolling and shopping are those that have a major anchor nearby or even anchors at both ends. In this sense, the street itself is very much like the malls we have become accustomed to. If the street is not too long (or wide) and if there are clear and desirable destinations at both ends, then it may indeed attract walkers. Sometimes even subtle differences in the spatial arrangement of things can mean success or failure in the fickle world of retailing. Having the pieces is one thing; putting them together properly is another.

Variable 5. Amount and Variety of Downtown Retail

Seattle: A (10)

Downtown Seattle has a nice mix of old and new big department stores, two modest enclosed malls, Pike Place Market, and a variety of street-level stores and cafes in specialized retail districts all around the downtown core. The flagship Nordstrom department store (along with a nearby Nordstrom Rack) anchors a consensus retail district at Westlake Park, which includes one of the few

new, freestanding department stores in an American downtown (now a Bon Marche). There are also three small urban shopping atria filled with specialty shops, cinemas, and restaurants. Pike Place Market, a huge public-market building dating from the early 1900s, contains a wide variety of crafts while maintaining its role as an authentic fish, fruit, and vegetable market. These two nodes are located only a few blocks apart and both have spun off a wide variety of street-level shops including everything from Indian arts and crafts to drug stores and T-shirt emporia. The waterfront promenade also houses a number of restaurants and tourist-oriented shops. Downtown has also become a "fun zone" for local teens as well as tourists, with a GameWorks, Nike store, and an REI "wilderness" store complete with a climbing wall. Small shops and cafes are becoming increasingly common around the edges of downtown in Belltown to the north and the International District and Pioneer Square to the south. The latter includes a large number of art and antique shops. Although many of the new stores are upscale, small magazine, liquor, and tobacco stores are holding out on many streets. Of course, Starbucks and Seattle Coffee shops abound.

Seattle's historic (1910) Pike Place Market

Portland: A— (9)

Portland has a very compact, consensus retail node at Pioneer Courthouse Square that includes a huge older department store (Mier and Frank), a new Nordstrom, and the two-square-block Pioneer Place Mall. In addition, the city requires street-level retailing on several downtown streets, and so there are a variety of specialty shops and traditional stores around the retail core. Beyond this core, however, the number and variety of stores diminishes rapidly. Still, there are restaurants and coffee shops on many streets and there is Powell's (huge) bookstore a few blocks away. Downtown has to compete for major retailing with the massive Lloyd Center Mall across the river. Lively Saturday markets help to create a sense of downtown excitement at least during the warmer months.

Indianapolis: A— (9)

Shopping in downtown Indianapolis is focused on the Circle Center Mall, located on a site that was identified as skid row in the 1950s. The mall is a large (million-plus-square-foot) facility anchored by Nordstrom and Parisian department stores that includes an exceptional variety of restaurants, cinemas, and entertainment venues. The outer walls were designed to either preserve or replicate the traditional city streets, and so it does not appear to be a big box. Linked to spin-off stores across the street by an aesthetically debatable "sky pod" that serves as an above-grade walkway, the mall has drawn a considerable amount of activity toward it. New street-level restaurants have also opened nearby. The mall has proven to be quite successful with local residents, who list shopping as the main reason for visiting downtown. Aside from the mall area, downtown has few vibrant retail streets. City Market, a traditional food market/food court at the other end of the core, is busy at noon, and Union Station will have specialty shops after renovation.

San Diego: A— (9)

Horton Plaza Shopping Center in downtown San Diego is one of the most famous and successful examples of whimsical postmodern design in America. The landmark structure, which opened in 1985 in the heart of the former skid row, made downtown San Diego a major shopping destination for the first time in three decades. Prior to the opening of the mall, the downtown had no department store and very little retailing at all. After considerable discussion and redesign, the

The Circle Center Shopping Mall in downtown Indianapolis has included many historic building facades.

mall was built to look like a jumble of open-to-the-sky structures that, theoretically, appeared to have accreted gradually over time. Bright colors and Mediterranean architectural details, combined with the preservation and integration of several historic structures (including a single-room-occupancy hotel), contribute to the center's whimsical character. The complex includes a variety of restaurants as well as a multiplex cinema and two live theaters. Nearby, a number of cafes and specialty shops have sprung up in the historic Gaslamp Quarter. Beyond this retail and entertainment zone, there is little shopping downtown, although a number of tourist shops exist at Seaport Village on the waterfront. There are no older department stores or specialized retail streets in other parts of downtown.

Pittsburgh: A− (9)

Downtown Pittsburgh has managed to remain a fairly important retail destination over the decades thanks to the huge, old Kaufmann's department store and the smaller, more recent additions of Lazarus and Saks Fifth Avenue. There are also a number of local, street-level shops downtown offering a wide variety of low- to high-end merchandise. Specialty shops are found in office building atria such as the Oxford Center and PPG Place. Station Square, across

Horton Plaza is a combination shopping center and festival marketplace in downtown San Diego.

Street-level shopping is still alive in downtown Pittsburgh.

the Monongahela River, functions as a festival marketplace with restaurants, crafts, and entertainment. The Strip, along the Allegheny River, includes traditional produce markets and a few interesting shops as well as some bars and restaurants. A major controversy in downtown Pittsburgh revolves around the plan to rebuild the traditional and lively but somewhat dowdy retail core along Fifth Avenue. Plans call for a mall-like street with more chain stores and cinemas. Preservationists oppose the plan.

San Antonio: A− (9)

Downtown San Antonio is a major tourist destination with tourist-oriented specialty shops and restaurants all along the famous Riverwalk and in other attractions such as Market Square and La Villita. Beyond the tourist zones, much of the street-level shopping seems a bit tired, although plans are afoot to revitalize it. The major downtown shopping venue is the ten-acre River Center Mall, which is linked to both the Riverwalk (and can be approached by boat) and to an older, freestanding Dillard's department store. The mall is also anchored by a Foley's and contains restaurants, cinemas, and entertainment venues. Tourist shops focused on the Alamo are located nearby. Some good lower-end retail still exists downtown as well.

The River Center Mall in San Antonio is connected to the Riverwalk.

Columbus: B+ (8)

For nearly a century, retail in downtown Columbus was focused on the eight-story, two-square-block Lazarus department store, at one time the fifth largest department store in the United States. In 1989, it was literally joined (by a second-level walkway) to the 1.4-million-square-foot City Center Mall and two new anchors—Jacobson's and Marshall Fields. After two decades of hanging on, downtown once more served as a major retail destination. Although connected to a hotel and performing arts center, the mall is light on fun, with no cinemas or entertainment venues. Blank walls predominate on the outside and connections with the urban fabric are not great, although there is room for expansion. Beyond the mall / Lazarus core, street-level retailing is minimal in the center of downtown. The only other retail destinations are the specialty shops in the Ohio (convention) Center and the art galleries and cafes of the Short North / North Market area at the northern edge of downtown. There are relatively few restaurants in the core, although there are many in the Short North and in the Brewery District to the south and many are opening in the new Arena District.

Baltimore: B+ (8)

Harborplace with its two-pavilion festival marketplace along the waterfront is very lively with street musicians, carts, and food stalls, but it mostly serves as a food court and as a tourist and specialty shop destination. The Gallery at Harborplace, a modest three-story vertical mall across the street, provides a wider range of shopping but has no major anchor stores. The Lexington Market on the northeastern edge of downtown offers fresh fish and produce and anchors a low-end retail district that adds diversity to downtown shopping. Antiques and crafts are found on the northern stretch of Charles Street and on Howard Street. Restaurants and small shops are also found in the nearby "Italian" district. Retail is expanding as the waterfront continues to deindustrialize. A Hard Rock Café and a Barnes and Noble now share space in a monumental former power plant. Downtown Baltimore is a fun place to shop and eat but is less important as a major retail destination.

Cleveland: B+ (8)

The Avenue at Tower City Center (in the Terminal Tower Complex) is the main retail destination in downtown Cleveland. The mall is tied to a traditional

large department store and includes the usual shops, restaurants, and cinemas as well as a water show and an enclosed tunnel to nearby sports venues and stairs to the lower-level commuter rail system. A few blocks away is a smaller, two-level mall called the Galleria. In addition, the Cleveland Arcade (1890) houses specialty shops in one of America's first arcades. Beyond these nodes, shops on traditional retail streets such as Euclid are struggling, except for some new sports bars. There are restaurants and bars in the Warehouse District and the Flats. Most downtown streets, however, are not too lively.

Minneapolis: B+ (8)

The retail district in downtown Minneapolis is focused on Nicollet Mall—a nicely designed pedestrian and bus mall in the heart of the city. The mall is linked to two enclosed shopping centers (City Center and Gaviidae Common) anchored by department stores such as Marshals, Daytons, Nieman Marcus, and Saks Fifth Avenue. The shopping centers are tied to a number of office buildings and hotels by a series of skyways. These second-level walkways also include a number of specialty shops and coffee outlets geared toward office workers. The skyway system and its associated retail make for pleasant "parka-free" shopping in the frigid Minneapolis winters, but they also make for an enclosed downtown with relatively little activity at street level. Also, the system can be a bit confusing. There is no obvious mall destination from the outside, only an internal maze. Beyond the retail core, specialty shops occupy Butler Square and other older buildings in the warehouse district along the river. There are also some shops and restaurants on nearby Hennepin Avenue.

Providence: B+ (8)

Even though it had one of the first shopping arcades in the nation (Weybosset, 1828), there was very little retail activity left in downtown Providence by the early 1990s. All of this changed beginning in 1997 with the opening of Providence Place Mall. With 1.3 million square feet of space and 350 stores, it is the largest mall in New England and matches the total amount of retail that existed in all of downtown during its heyday. Anchored by Filenes, Nordstrom, and Lord and Taylor's, the center also contains restaurants, cinemas, and meeting space. The design, after much discussion, relates well to the city with the eastern (city) side designed to look like a series of "colonial" buildings rather than a big box. Large windows let shoppers and diners look out on the city skyline and riverfront park. It is located in what was dead space between the down-

Providence Place Mall (the largest in New England) sits between the state capitol and the downtown core.

town core and the hilltop state capitol building. Beyond the mall, there is little serious shopping in downtown Providence. The arcade contains specialty shops, as do many of the small, picturesque streets, but some of these may suffer as the mall fills. There are no remaining older department stores. With a next-door Amtrack station, Providence Place Mall could become a shopping destination from as far away as suburban Boston.

St. Louis: B (7)

Downtown St. Louis has one of the largest festival marketplaces in the world, housed in the former Union Station. The station itself was once the nation's largest railroad terminal and today it contains shops, restaurants, music venues, a hotel, and even a lake (as well as lots of covered parking), all under the rehabilitated train shed. Nearby, a multiplex cinema takes advantage of the symbiotic relationship. However, the entire area is geared more toward fun than serious shopping. There are also some specialty shops in the historic Laclede's Landing district along the Mississippi River, but restaurants and bars predominate. Traditional downtown retailing is more problematic. The St. Louis Center, a four-story glass galleria, connects two large department stores (Dillard's and

Famous-Barr) in the heart of the CBD but doesn't seem to be attracting throngs of shoppers. Some street-level retailing hangs on but with no real foci.

Atlanta: B (7)

Even though it has a number of retail outlets, downtown Atlanta doesn't quite work as a major shopping destination. Many of the upscale specialty shops in Peachtree Center and the major hotels are enclosed in fortresslike structures and geared as much toward convention goers as locals. The more traditional stores around Five Points have either closed or are in marginal condition. The once vibrant tourist attraction at Underground Atlanta contains specialty shops and restaurants but has experienced hard times recently (again), and many of the street-level shops have closed. The only downtown department store, Macy's, may well close in the near future. The emphasis on megastructures, especially in the newer, northern section of downtown, makes it difficult to develop "friendly" street-level retailing. Midtown shops and Lenox Mall in Buckhead, a few subway stops away, draw away many potential downtown shoppers.

Denver: B− (6)

The Sixteenth Street pedestrian and bus mall is as lively and attractive as any retail core around. There are trees, benches, boutiques, cafes, and entertainment venues. The only thing it lacks is serious shopping. While there are plenty of dining, dancing, and specialty shopping opportunities, especially at the new Denver Pavilions, there are no major department stores or large traditional shopping centers. The largest venue is the Tabor Center, a glass-enclosed galleria with seventy shops. There are also some specialty and antique shops (and a large bookstore) in the historic Larimer Square and LoDo districts, but sports bars predominate.

Phoenix: C− (3)

Aside from a few small specialty shops in settings such as the Arizona Center, downtown Phoenix has almost no retailing. There are some sports bars, a multiplex cinema, and some restaurants. There are no major department stores, malls, or traditional shopping streets in the downtown area. Even in Midtown, a few miles away on the Central Avenue spine, an older mall has closed. Downtown Tempe and Scottsdale seem to attract those who want to stroll in a retail environment.

Charlotte: C— (3)

Downtown Charlotte is not a retail destination. Aside from a few (twenty) small shops in the atrium in the NationsBank Tower and an antique center called the Firehouse, there is almost no place to shop. A small, historic arcade has been renovated but there is little retailing there. Restaurants and bars are few and far between as well, and there are no support streets or districts on the fringes of downtown.

Summary: Downtown Retail

Seattle 10
Portland 9
Indianapolis 9
San Diego 9
Pittsburgh 9
San Antonio 9
Columbus 8
Baltimore 8
Cleveland 8
Minneapolis 8
Providence 8
St. Louis 7
Atlanta 7
Denver 6
Phoenix 3
Charlotte 3

The Visitor Industry: Hotels, Motels, and Convention Centers

Throughout urban history, travelers have stayed overnight in a variety of inns and lodgings of varying quality. The modern "full service" hotel came into being during the 1820s with Boston and New York City leading the way. The large hotel with dining rooms, elegant lobbies, and a full range of services and shops evolved during the nineteenth century along with rail travel. Many of the very largest hotels in Canada were built by the railroads themselves, while in the United States they were often simply close to major stations. Rail travel also made it possible to build major resorts with massive seaside or mountain

"lodges." With improvements in intra-urban transportation, including horse-cars, streetcars, and later motorized taxis, hotels were built further away from the often smoky and grimy train depot area, and as close as possible to the most prestigious locations.

By the turn of the twentieth century, some hotels were becoming very large and opulent indeed. With the advent of elevators, indoor plumbing, electric lights, and other "modern" conveniences, a stay in a first-class hotel was like a trip to the future. People could experience amenities that they were not likely to have in their homes. Elegant hotels were dubbed "people's palaces" since the lobbies and dining rooms achieved a scale previously known only to royalty, or at least the very rich. To a degree, this tradition continues as major hotels compete to create elegant and unusual settings in order to attract high-paying guests.

Of course, not everyone could afford to stay at such "palaces," and so other hotel traditions evolved as well. Small, budget hotels with the bathroom down the hall appeared on the fringes of downtown, and some of the older, nicer hotels near the train stations joined this category as newer, grander places opened up. Also, there were still a number of signs advertising "rooms" in everything from private houses to the spaces above shops.

By the 1950s, there were three distinctive hotel options in and around the typical American downtown. At the top of the hierarchy were the huge, first-class downtown hotels. Some of these had more than a thousand rooms as well as a full range of luxurious facilities and services, including doormen in outlandish costumes. As in the case of department stores, the earliest hotels were usually locally owned and controlled, but gradually hotel chains such as Hilton and Sheraton became dominant. The second type of hotel—although distinctions were sometimes a little blurry—was the smaller hotel with few services. These relied on nearby cafes, shoe repair shops, beauty parlors, and other such conveniences to help keep their guests happy. Both of these types of hotels served as long-term residences as well as short-stay accommodations. Hotel living was thought to be quite proper until at least the 1920s.

The third variation on the theme was the newly "invented" motel. These, at first in the form of "cozy cottages," were usually found in suburban or rural locations where land was cheap, but they could also be found on the edge of downtowns. The important thing was that you could park in front of your room, avoiding the formalities associated with a lobby. Over the years, each of these three types of accommodations has experienced ebbs and flows in popularity.

Downtown Hotels in the Postwar Years:
The Search for a Style and a Location

Downtown hotels, like a lot of other downtown functions, fell on hard times during the 1950s and 1960s. Many of the smaller, older hotels, especially those with large numbers of long-term residences, were categorized as "skid row" dwellings and slated for urban renewal. The term *transient* became a very negative designation, and old hotels became synonymous with downtown sleaze in the eyes of many planners and social reformers. Thousands of single-room-occupancy hotels (SROs) have been destroyed in U.S. downtowns since the 1950s.[3]

Even large, luxury hotels had problems as America entered the automobile age and largely abandoned travel by train. By the late 1950s, nearly everyone needed a place to park, and few downtown hotels had many such spaces available. Some were able to rent space in nearby abandoned warehouses and offer valet service to take cars there. Others were able to knock down adjacent smaller structures and build parking garages, but this was very difficult to accomplish in the heart of the CBD. In addition, many in the "rock and roll" generation viewed the downtown luxury hotel with its formality and ostentation as a bit passé, especially since the decor was often looking a little tired by the late 1960s. It did not help that the rooms and closets were small and the plumbing facilities awkward compared to those in the new, sprawling roadside motels. The latter sometimes even had swimming pools.

As new highways were built and old streets widened in and around American downtowns, most visitors came to think of motels as the most normal type of urban accommodation. The result was a spreading out of the visitor industry. Even people hoping to visit the shopping and cultural attractions of downtown increasingly cruised the strip and stopped for the night at least a mile or so from the city center. Street life diminished and the impact of tourism was diluted as folks ate at roadside diners. Over the decades, urban motels tended to get bigger as multistory hundred-plus-room establishments pushed out the cozy cottages of earlier years. These "motor hotels" sometimes included restaurants, swimming pools, and, today, even fitness and office centers. The line between the hotel and motel has become increasingly fuzzy. In relatively low-density downtowns with lots of nearby space such as San Antonio and Indianapolis, motor hotels play a very important role in the visitor industry. Families traveling by car seek out nice, centrally located but informal establishments with adequate free parking.

In the meantime, many of the older, smaller motels have experienced a fate very much like what the older railroad hotels went through decades earlier. If they still exist at all, they are likely to be rented by the week or month (or hour) to people who are not usually classified as tourists or even visitors. This makes it difficult to decide just what to include in the count of downtown hotel rooms.

The Return of the Downtown Luxury Hotel

During the 1960s, there were some halting steps taken in a variety of downtowns aimed at bringing back the huge, luxury hotel. Most of these were modest places by today's standards and resembled monster versions of the motor hotel as much as the classic downtown hotels of the interwar years. They did, however, normally have in-house garage parking, restaurants, and amenities such as swimming pools. Above all, however, they had meeting space. Large organizations still needed places to meet, and downtown hotels eagerly sought to capture a share of this booming market. Well-equipped meeting rooms and ballrooms became a very important part of new hotel design.

The trick was to provide something downtown that was novel and attractive as well as functional. If the old hotels seemed formal and stuffy, what new designs might be used to create modern settings with enough glitz and pizzazz to bring people back downtown? The answer, according to architect/developer John D. Portman of Atlanta, was a cavernous atrium. The 1970 opening of the Hyatt Regency on Atlanta's Peachtree Street upped the ante for those seeking to build monumental downtown hotels. The Hyatt has a twenty-two-story atrium lobby with cafes and ponds at ground level. A protruding blue-domed revolving restaurant tops the building. In addition to being the first "atrium hotel in the world," it was also the first of the modern, self-contained megahotels. It has 1,264 rooms, contains plentiful meeting and restaurant space, and is linked to shopping and office complexes. The hotel was something to see in the early 1970s, and it was soon followed by variations on the theme in San Francisco, New York, and other cities.

By the early 1990s, nearly every major American downtown had at least one and usually several modern megahotels with five hundred to a thousand rooms and a wide range of in-house facilities. The market is dominated by a few chains such as Hyatt Regency, Marriott, Hilton, Omni, Westin, and Adams Mark. Most of the hotels are new, but a few older ones have been renovated and enlarged so as to mix the charm of the past with the current need for modern conveniences. There are three types of locations favored by the new megahotels: the

traditional "main square" setting in the heart of the CBD; along newly designed waterfront parks and promenades; and in association with major convention facilities. Depending on the spatial arrangement of these kinds of attractions in particular cities, the hotels may be highly concentrated or widely dispersed within the downtown.

In order to capture every possible segment of the tourist and business market, cities have encouraged the construction of other types of hotels and motels as well. Holiday Inns, Days Inns, and Comfort Suites typically serve a slightly less affluent (or nonconvention) clientele while Embassy Suites, Courtyard by Marriott, and others seek to offer a slightly different product. Small, renovated historic hotels and bed and breakfasts have also appeared in and around most successful downtowns. A few cities have fought to preserve existing SROs and even to build new ones, and pensions and youth hostels, long popular in European cities, are popping up as well.

Downtown as Meeting Place: Convention Centers, Trade Shows, and Exhibit Halls

It seems ironic in the age of electronic communications—telephones, faxes, email, the World Wide Web, and two-hundred-channel television—that face-to-face meetings in the form of major conventions are more important than ever. It also seems ironic that in an age when downtown land values have skyrocketed and the size and variety of stuff being displayed has increased, exhibit halls have moved inward toward the city center rather than outward toward cheaper land. Downtown hotels typically book one large professional group after another in an endless stream of conferences, banquets, parties, and special events. Many of these events have always taken place in major hotels, but huge trade shows are a relatively new thing in most American downtowns.

Giant trade shows in the form of world fairs first appeared during the middle of the nineteenth century in Europe and the United States. Host cities usually built elaborate new facilities for these shows, such as the Crystal Palace in London, but none of them was downtown. Most were built in parks or suburban locations such as Forest Park in St. Louis, Hyde Park in Chicago, or the Marina District in San Francisco. For smaller trade shows and annual events, most communities created permanent fairgrounds complete with exhibit halls and a coliseum. Horse shows, tractor shows, car shows, boat shows, and, of course, annual fairs, were all typically held at the fairgrounds. Even large political con-

ventions were often held in peripheral exhibit halls such as the Cow Palace on the southern fringe of San Francisco.

In 1950, very few cities had large downtown convention centers and even fewer had such facilities in the heart of the CBD. Slum clearance and new civic centers associated with the City Beautiful Movement cleared the way for a few convention complexes, such as the one in Cleveland, but these were the exception rather than the rule. Most exhibit halls were on the edge of downtown, like McCormick Place in Chicago, if they were downtown at all.

During the 1950s and 1960s, most cities built some kind of civic auditorium or sports arena that could hold large downtown gatherings, and many of these included some exhibit hall space, often in the basement, suitable for small trade shows. Most also had a large number of meeting rooms with moveable walls so as to accommodate different kinds of events. The full-blown modern downtown convention center, however, appeared in most cities only during the 1970s and 1980s. The earliest versions were modest and somewhat sterile boxes with perhaps a hundred thousand square feet of more or less horizontal and contiguous space. Most of these have been either greatly expanded or replaced over the past three decades. Today, most mid-sized American downtowns have convention centers with between two to six hundred thousand square feet of exhibit space. A few centers, usually on the periphery rather than in the center of downtown, have over two million square feet of total gross space. For example, McCormick Place, on the periphery of downtown Chicago, contains 2,600,000 total square feet.

Three very important questions loom large in discussions about the future of downtown convention centers: how big can they be, where should they be, and what should they look like? Given the fact that convention space needs to be largely horizontal, there is a limit to how large downtown centers can be, especially if they require huge areas for parking, loading docks, and the like. The temptation to build bigger and better centers must be combined with sensitivity to how they relate to the smaller-scale activities around them.

This brings us to the second question, just where within a downtown should large convention centers be located? When huge boxes are plopped down right in the middle of the CBD, the result can be vast blank walls and urban dead space, especially since streets must be closed for superblocks in order to accommodate such large structures. On the other hand, if centers are sited too far away from the center of downtown, they may not contribute much to the life of the city or contribute to the business of downtown hotels and restau-

rants. If large convention centers are located on the waterfront, they may attract business on the basis of amenity value, but they may also wall off the attraction from the rest of the city. The best locations, it would seem, are right
on the edge of the downtown core, perhaps over a freeway (Seattle), abandoned
railroad tracks (Columbus), or close to, but not on, a waterfront (Baltimore).

Finally, how can big boxes be designed to be interesting architecturally and
to add at least some fine-grained detail and life to the city outside as well as
functional space inside? Some progress is being made, although attempts at architectural innovation on the scale of a massive convention center are almost
bound to be controversial. In San Diego, the bayside center features a canopy
of sails in an attempt to reference the sailboats nearby. In Columbus, the center is designed to look like several different-colored small boxes instead of one
big one. In San Antonio, the Riverwalk is being expanded to flow through part
of the convention facility. Still, combining good design with a one-million-
square-foot box is a tough task.

A final and related issue concerns the hotel and convention center as fortress
rather than as an integral part of the downtown fabric. In too many cities, it is
possible for a visitor to spend five days at a convention without ever leaving the
hotel-shopping-meeting rooms-fitness center-restaurant-night club-parking
garage-hair salon megastructure. If hotels and convention centers are to contribute in any meaningful way to the overall quality of the downtown, private
spaces and public spaces must interact. After all, downtown hotels are typically
full of people who would like to go out and experience what cities have to offer in
terms of food, music, and culture. They are either on vacation or meeting friends
and colleagues at conferences. In this sense, each hotel resident may have more
impact than a permanent downtown resident who is busy with everyday life.

The Hotel as Theme Park

As downtown hotels have gotten larger and more elaborate, it is tempting to
think about what might happen if developers decide to go all out as they have
done in Las Vegas. The monster hotel as theme park is epitomized by Las
Vegas hotels such as New York–New York, Bellagio, Caesars Palace, Luxor, and
Excaliber. Las Vegas has well over a hundred thousand hotel rooms, mostly
located along the three-mile strip. The largest, the MGM Grand (billed as the
world's largest hotel) has 5,005 rooms—more than all but a dozen or so American downtowns have. Gambling and entertainment have traditionally been
the main attractions, but increasingly, many people go just to see the hotels.

Battles between pirate ships, a ride up the Eiffel Tower, a visit to the Egyptian desert, and lunch at a New York deli can all be accomplished in the space of a few blocks. The question arises: will hotels in American downtowns try to build on the local sense of place or will they increasingly become pleasant but place-less oases totally unrelated to the landscapes around them?

Hotels as destinations, it would seem, are likely to become more important as many large professional organizations that meet regularly, but do not bring products to display, seek space in hotel meeting rooms rather than renting extra space in convention centers. By booking a certain number of hotel rooms, organizations can get better deals on in-house meeting space. This usually means finding megahotels with around a thousand rooms and lots of amenities. To justify such an investment, developers must make sure that the hotel is both extremely well located and that it will become an important and memorable architectural landmark.

Counting Downtown Hotels and Measuring Their Impact

Several problems arise in trying to count the number of downtown hotels and hotel rooms. The two that are most important are the questions of where downtown is and what a hotel is. There are many sources of data, including tourist brochures, city sites on the web, AAA Tour Books, individual hotel chains, and, of course, field work. Nearly every source either misses some hotels or includes several that are not really downtown. In most cases, the total number of hotel rooms is underreported, since a large number of older hotels and motels are now used as semipermanent housing as much as for visitors, and many cities view them as an embarrassment. Small pensions and youth hostels are usually not mentioned except in specialized publications.

There is also the old problem of defining downtown. I have identified several "downtown support zones" that include motels, bed and breakfasts, and small hotels that are close to the downtown core but not really part of it. One of the big differences between American downtowns is that some cities, like Seattle and San Antonio, have many hotels in such nearby support areas while others, such as Cleveland and Pittsburgh, have few or none. In the latter downtowns, motels are nearly absent since they cannot easily locate in the CBD core and there are few desirable options. Still, I have tried to come up with some reasonable approximations of the number of hotels and convention facilities in the selected downtowns in the year 2000.

The number of hotels and hotel rooms in and around downtown is a pretty

good measure of downtown viability. There are certain cut-off points that are fairly consistent, at least in cities that do not have an exceptionally large tourist industry. For example, second-tier cities such as Indianapolis, Columbus, and Portland tend to have ten to twelve downtown hotels with at least three thousand rooms. Cities that are only slightly smaller and further down the urban hierarchy such as Toledo, Dayton, Fresno, and Omaha tend to have only two or three hotels and usually less than one thousand (and often less than five hundred) rooms. Of course, in cities with major tourist attractions such as San Antonio, the numbers in both categories can be much higher. Hotels do not tell the whole story, but they are a good indication of where significant numbers of people want to spend some time. Things tend to snowball. Downtowns that have lots of people on the street tend to attract more hotels full of people who want to enjoy the street scene.

Variable 6. Downtown Hotels and Convention Centers

Seattle: A (10)

There are about thirty-nine hotels, with a total of 7,412 rooms, in the greater downtown Seattle area, supporting a large and varied tourist industry. Hotels are located throughout downtown, from Pioneer Square in the south to the Seattle Center in the north. There are twenty-six hotels in the CBD core, close to the Washington State Convention Center and major cultural and shopping attractions. There are also many hotels and motels close to the Space Needle and other attractions at the Seattle Center. Hotels range in size and price from the thirty-five-story, 840-room Sheraton Hotel and Towers to a variety of small hotels and bed and breakfast (B and B) establishments. At the present time, there are still areas close to the core that are attractive for low-density motels with free parking, which makes the downtown more attractive for budget tourists. The 200,000-square-foot convention center, famous for being built over a freeway, is being expanded and linked to the new History and Science Museum. The main entrance to the center on Pike Street faces a lively street and is only two blocks from the retail core. Waterfront motels and meeting space are beginning to appear in order to take advantage of an obvious amenity.

San Antonio: A (10)

Downtown San Antonio with its Riverwalk and Hemisfair Plaza is a major urban tourist attraction emphasizing the simple pleasures of food, music,

strolling, and an exotic ambience. Major attractions are well integrated into the overall city theme. There are thirty-one hotels, totaling 8,031 rooms, in the greater downtown area, with twenty-five in the core. Hotels range from huge, luxury complexes such as the thirty-story, thousand-room Marriott Rivercenter to small bed and breakfasts and budget motels. The formerly light industrial region south of the downtown core is being colonized as a motel district, and the King William Historic District provides a location for small B and Bs. The convention center, located in the former world's fair complex (Hemisfair) is being expanded and linked to the Riverwalk.

San Diego: A (10)

Downtown San Diego is situated between San Diego Bay and Balboa Park and is a major urban tourist attraction, even though most tourists have traditionally gone to beach communities rather than the city center. There are at least twenty downtown hotels, providing about 6,400 rooms, with most of them located either along the bayside promenade or a few blocks away. The waterfront convention center currently has about 350,000 square feet of exhibit space but is being expanded to include nearly 600,000 square feet. The center is controversial in that it blocks the approach from downtown to the waterfront, although there is public space along the bay. There are another two-thousand-plus rooms located close enough to have views of the downtown skyline and easy access to downtown attractions in nearby support districts such as Coronado and Harbor Island. Hotels range in size and price from the huge Marriott Hotel and Marina with 1,354 rooms to small pensions, budget motels, and historic B and Bs. The prospects for more hotels in and around downtown should be good as the eastern part of downtown is developed. Still, the convention industry argues that the downtown is short of first-class rooms.

Atlanta: A (10)

Atlanta is a major air hub and central place for the Southeast and so functions as a very important place for a variety of professional meetings and conventions. Downtown Atlanta thrives on business more than recreational tourism, although the latter has increased with major events such as the Olympic Games. The Georgia World Congress Center is one of the nation's largest convention halls, with over one and a half million square feet of exhibit space, and other venues such as the Georgia Dome and in-house hotel space are also important. The core of downtown Atlanta has fifteen hotels, with 6,757

The San Diego Convention Center and associated hotel towers

rooms, heavily concentrated around the Peachtree and International Boulevard core. Downtown is known for massive hotels with large amounts of self-contained space. There are three hotels with more than one thousand rooms, including the Westin, which is the tallest hotel in America at seventy-three stories. The megahotel as self-contained amenity entered the modern age during the early 1970s with the John Portman–designed Hyatt Regency with its twenty-two-story atrium and 1,264 rooms. Outside the megastructures, however, meeting space and hotels are not well linked. There are no hotel support districts near downtown Atlanta, and activity falls off rapidly away from the core, although a more humanely scaled hotel district exists in the Midtown-Buckhead area a few miles to the north.

Baltimore: A− (9)

As Baltimore has become part of the greater Washington, D.C./Baltimore CMSA, downtown Baltimore has tended to specialize in tourism and recreation while Washington functions as the major center for business. The once-grimy waterfront has emerged as a major tourist destination over the past twenty years and is now considered to be an ideal place for lunch and a ballgame and perhaps

The northern part of downtown Atlanta is made up of a series of giant and interconnected convention hotels.

a visit to a museum. The convention center contains 300,000 square feet of space and is located between the shops at Harborplace and the ballpark at Camden Yards. There are sixteen hotels, with 4,553 rooms, in the downtown/harbor area, with more on the horizon; but daytrippers from within the heavily populated region make downtown hotels less vital than in cities like San Antonio. The largest hotel is the Omni Inner Harbor, with 707 rooms. It is located in the harbor and convention core. There are few supporting hotel districts, although Fells Point, a mile to the east along the waterfront, is a historic attraction.

Denver: A− (9)

Downtown Denver has become a visitor destination in its own right as well as being a gateway for tourists on their way to the Rockies. Although it is now

The Denver Convention Center and the adjacent performing arts center

a major air hub, it remains the center of a lightly populated region and so it is not as important a business center as, say, Atlanta. Downtown Denver has fifteen hotels (with 5,169 rooms) scattered throughout the downtown area from the state capitol to the historic LoDo district. Hotels include the historic Brown Palace and the huge Adams Mark (1,225 rooms) as well as some small bed and breakfast establishments. The convention center contains over 300,000 square feet of exhibit space but will double in size by 2004. It consists of two structures on the edge of the downtown core next to the performing arts center, thus making a wide variety of "gathering" space available. The backs of these buildings, facing Cherry Creek and the Educational Institute of the Auraria campus, are bleak, but the fronts facing downtown are well designed.

St. Louis: B+ (8)

Downtown St. Louis has eleven hotels, with 4,484 rooms, mostly located close to the Gateway Arch along the Mississippi River and the convention center, with a secondary center located around the renovated Union Station. The huge convention center (over 500,000 square feet) is combined with the enclosed football stadium (TWA Dome) to form a megastructure of monumen-

The St. Louis Convention Center complex includes a domed football field.

tal proportions that essentially takes up much of the northern side of downtown. The hotels include the twenty-eight-story, 780-room Regal Riverfront and a 538-room Hyatt Regency located in the former Union Station. Linkages are good since the hotels around the convention center are close to the restaurants at Laclede's Landing, the waterfront park with its riverboat gambling venues, and Busch Stadium. The Union Station node seems a bit separate from the CBD core, but it is linked by light rail. There are few hotel support zones near the downtown core, since much of the central city has experienced severe disinvestment. A motel zone, therefore, is difficult to establish.

Portland: B (7)

Downtown Portland consists of two separate and very distinctive hotel and convention districts. The main convention center (300,000 square feet) is lo-

cated across the Willamette River from the downtown core and is thus peripheral to it, even though the two areas are connected by light rail. The two districts combined have twenty-three hotels, totaling 4,405 rooms, but only fifteen hotels (with 3,550 rooms) are in the downtown proper. The Marriott is the largest hotel but has only 503 rooms, and so it may not be big enough for major conferences, although several hotels are located in close proximity. Future hotel and convention expansion will probably be across the river in what might best be classified as a downtown support zone.

Minneapolis: B (7)

Downtown Minneapolis functions as a major central place for the Upper Midwest. It has thirteen hotels, with 4,114 rooms, the largest being the 821-room Hilton located near the expanded (500,000 square feet) convention center. Historic options on the edge of downtown include hotels in a former flour mill along the Mississippi River and a converted factory on Nicollet Island. The convention center is located on the southern flank of downtown but is linked to a variety of hotels and shopping centers by the skyway system. The system helps to keep the downtown core compact. The fact that most of the major hotels are linked helps to provide a threshold population for core area shopping and dining.

Indianapolis: B (7)

In that Indianapolis is a state capital and an emerging center for amateur and professional sports, hotel and meeting space is increasingly in demand. The downtown area has sixteen hotels, with 3,616 rooms, many of them having been built in the last decade. To date there are no huge hotels and the largest is the five-hundred-room Hyatt Regency. Most of the hotels are clustered around the megastructure that includes both the 400,000-square-foot convention center and the RCA Dome (football stadium). A hotel is also located in the renovated Union Station. There are a few nearby hotel support zones, and additional space is still available along the redesigned White River. Many nice motels continue to be located downtown.

Cleveland: B− (6)

Throughout most of its history, downtown Cleveland has not been associated with a tourist industry, although it was one of the first downtowns to have a massive convention facility. The city is trying to make up for lost time, but hotel options are still relatively limited considering the boom in sports venues

and museums downtown. The compact downtown has ten hotels, with 2,815 rooms. More are under construction and will soon push the total past the three thousand mark. There are as yet no megahotels, with the largest having only 491 rooms, thus limiting the number of large conventions. The convention center dates from the 1920s but has been modernized and expanded. It is very well located and relatively picturesque for a "big box." Plans call, of course, for a newer and larger facility. One reason for the relatively small number of downtown hotels, given the large size of the metropolitan area, is the lack of suitable sites for motel support zones. Industrial landscapes along the Cuyahoga River and urban decay on the near-east side limit the possibilities for hotel construction in neighborhoods close to downtown.

Columbus: B− (6)

Columbus is the centrally located capital of a populous state and it hosts a fairly large number of regional conventions. Major exhibits, meetings, and sports events were traditionally held at the fairgrounds, but today the downtown convention, hotel, and sports complex business is going strong. The unusual, and sometimes controversial, convention center designed by Eisenmann

The Columbus Convention Center attempts to avoid the big-box look with a jumble of forms and shapes.

is currently being expanded to 450,000 square feet of exhibit space, while other downtown venues such as Veterans Memorial can also be used for meetings. There are about ten downtown hotels, with 2,661 rooms, and one more set to open. The largest, the Hyatt Regency at the convention center, has 631 rooms. Two of the hotels are in renovated historic structures. The hotels are spread throughout downtown with nodes at Capitol Square and near the Brewery District as well as at the convention center, and activity can be spread a bit thin. At present there are few supporting districts, but small hotels and bed and breakfasts do exist in nearby German Village and the Short North.

Pittsburgh: C+ (5)

In the days of steel and smoke few people wanted to stay anywhere near downtown Pittsburgh unless they had to. Those days are over now and the city has experienced a renaissance. Even with the boom in office space and waterfront revitalization, however, the city has been slow to develop a downtown tourist industry and first-class convention facilities. There are only six downtown hotels, with a total of 2,338 rooms, and only four are in the compact CBD core. The largest, the historic William Penn, now a Hilton, has 712 rooms. Across the Monongahela River, the Sheraton occupies a site at Station Square,

The giant William Penn Hotel still occupies a central square in downtown Pittsburgh.

a river- and rail-oriented festival marketplace. As in Cleveland, the physical site and the industrial character of the landscape has limited hotel construction on the fringes of downtown, and so motel districts are absent. The city is rebuilding the rather small and dreary convention center with better linkages to the now attractive Allegheny River. Future hotels might appear north of the river in association with new sports complexes.

Charlotte: C+ (5)

Charlotte is a city on the make and has boomed in recent years. With a major airport, a stunning skyline, and a large, if sterile, 400,000-square-foot convention center, downtown Charlotte seems primed to try to take some of the burden off nearby Atlanta. Downtown has eight hotels, totaling 1,954 rooms, with more coming. So far, however, the largest hotel has only 434 rooms and there are few hotel support districts around downtown.

Providence: C (4)

The downtown Providence hotel industry had a near death experience but is coming back. There are now four hotels in the downtown area, with a total

A convention center, civic arena, and high-rise hotel provide a clear edge for downtown Providence.

of 995 rooms. The small but centrally located convention center is nicely linked to two of the hotels, especially the twenty-five-story, 365-room Westin. The city suffers as a tourist and business destination due to its proximity to Boston. It is also the center of a state that is barely larger than the metropolitan area itself. Hotel support districts could emerge in nearby historic neighborhoods and along the harbor.

Phoenix: C (4)

Phoenix has far fewer downtown hotel rooms than one would expect in a major American metropolitan area. There are four hotels in the greater downtown area, with a total of 1,398 rooms. Only two of these, with a combined 1,244 rooms, are in the CBD core. They are adjacent to each other, however, and to the 300,000-plus-square-foot convention center and sports venues, thus permitting moderately large gatherings. There is lots of room for motel support districts, but most of the tourist activity is focused on Scottsdale, although some motels are located further north on the Central Avenue midtown spine.

Summary: Hotel and Convention Space

Seattle 10
San Antonio 10
San Diego 10
Atlanta 10
Baltimore 9
Denver 9
St. Louis 8
Portland 7
Minneapolis 7
Indianapolis 7
Cleveland 6
Columbus 6
Pittsburgh 5
Charlotte 5
Providence 4
Phoenix 4

Downtown Expands

Major Attractions, Historic Districts, Residential
Neighborhoods, and Transportation Innovations

Downtowns are no longer simply central business districts. Most viable center cities now encompass at least a square mile or two and include a variety of support districts for the traditional activities discussed in the previous chapter. Among these are major entertainment and cultural attractions, historic districts and related light industrial zones, and downtown and near-downtown residential neighborhoods. Most of these districts are space-extensive compared to the traditional, compact central business district, and so the spatial arrangement of things has become increasingly important. Not only do good downtowns require a wide variety of new and old activities; these activities must be located so that the whole is greater than the sum of the parts. They must be seen as contributing to downtown as one place. Therefore, I have chosen to include in this chapter attempts to create transportation innovations, beyond the automobile, that can help to link downtown activities together.

Major Attractions: Sports, Culture, and Entertainment

Admittedly, the term *major attractions* is a bit fuzzy. After all, many of the traditional functions or site characteristics discussed in chapters 3 and 4, such

as department stores, office skyscrapers, and waterfront promenades, could also be described as major attractions. The historic districts that will be discussed later in this chapter might also be classified as major attractions. For the purposes of this section, however, the term will be used primarily to describe sports venues, museums and cultural centers, theater and performing arts facilities, and a few miscellaneous semidowntown activities such as zoos and harbor excursions. Most good downtowns are currently experiencing a reconcentration of these activities as the move toward creating "fun zones" in the central city gains momentum.

Tourism is the fastest-growing industry in the world, and American downtowns are eager to cash in on the trend by embellishing the central business district with a "city as theme park" ambience. This trend has been criticized on several fronts. First, sports venues and cultural centers are seen as catering to an affluent, often suburban clientele while ignoring the needs of nearby neighborhoods. Second, sports facilities are often condemned as financial boondoggles that use city tax money to subsidize wealthy team owners. Third, it is argued that sports facilities do not contribute to overall economic development in a district but may actually have a negative influence. Most stadiums make economic sense only if they are used regularly for a variety of activities including concerts, rodeos, motorcycle racing, and the like.

A major variable in these discussions is the degree to which major attractions act as catalysts for nearby revitalization. In some cases, sports arenas, performing arts centers, and museums have resulted in new life downtown as cafes, restaurants, shops, and even housing have been built as close as possible to the action. A disturbing trend, however, is the increasing tendency for ballparks and even museums to be self-contained "theme parks" with their own restaurants, clubs, auditoriums, shops, and related (or even unrelated) entertainment centers. The same could be said of amusement parks, aquariums, and other major tourist destinations. Such complexes could hurt surrounding businesses since there is no reason to venture beyond the gates of the entertainment center. Using tax money to subsidize such fortresses is obviously controversial.

While I take no stand on whether expensive downtown facilities are good in a moral, financial, or cosmic sense, I do argue that they can help to create what I define as a realistically good downtown. Even if they contribute to a city divided into "haves" and "have-nots" (a hotly debated topic), this may be better than a city with only "have-nots." Pride and hope can go a long way in putting cities back together. The challenge is to build sports arenas and cultural attractions in an intelligent and responsible way.

Sports Venues: Neighborhood Ballparks

For more than a hundred years, American cities have strongly identified with the sports teams bearing their names. For many, pride of place rises and falls with the league standings. Sport is thus a major entertainment attraction, but it is also much more than that. The big four of (men's) baseball, football, basketball, and hockey dominate professional sport in American cities, although soccer, women's basketball, and a few other teams are gaining popularity. The major league teams' venues have had by far the biggest impact on downtowns, but minor league stadiums and college arenas have also occasionally contributed to the mania. Of all these sports, professional baseball has been around by far the longest—over a century. The others have become important only since World War II.

It is hard to discuss the location decisions involved in siting major sports facilities, since they are usually the result of a multitude of actors and a long string of compromises and deals. Some teams have built their own arenas while others have used public facilities. In recent years, most sports venues have resulted from complex mixtures of public and private financing, urban renewal policies, and a wide array of rental arrangements. Even though it is not always easy to determine the "agents" involved, it is possible to trace the changing intra-urban locations of sports facilities over time.[1]

In general, the earliest locations for professional baseball teams were central city neighborhoods, usually along streetcar lines heading out from downtown. The parks once used by the Brooklyn Dodgers, New York Yankees, Pittsburgh Pirates, Chicago Cubs, Cincinnati Reds, Baltimore Orioles, and many other early teams epitomized this pattern. The organization (most early ballparks were privately financed) sought the best, most accessible location they could afford, but since sports stadiums took up a lot of space compared to most downtown activities, the norm was to be well away from the peak land value intersection. This was especially true in the years before jet travel, when most of the teams were located in the compact cities of the Northeast and Midwest. Regular trips to, say, California or Florida were impossible to work into the schedule. For a half-century, St. Louis was the end of the known world. When professional football became popular, a city's two sports teams often shared facilities, since few teams could afford to build their own quarters until the 1960s. Often local teams rented municipal stadiums.

Professional hockey (1927) and basketball (1947) became major attractions

long after baseball had paved the way. These teams could usually be located closer to the central business district since they occupied smaller (15–20,000 capacity), or at least more vertical buildings. Centrally located venues such as Madison Square Garden in New York City and Boston Garden quickly became famous venues for "inside" sporting events. (It is interesting that these places were called gardens, but that's another story.) During the early years, however, even these sports were more likely to be played in a coliseum at the fairgrounds or some other "neighborhood" location than in downtown.

In general, these neighborhood locations are currently the least popular for new sports venues. Only major league baseball still has a significant percentage (about 40%) of its facilities in neighborhood locations, and this is largely a holdover from the past. Over the decades, most teams have either moved outward to a new and spacious suburban location or inward toward a revitalized downtown. Until quite recently, the former move was most common. During the 1960s and 1970s, many teams were moving to relatively remote "outer belt" locations such as Foxboro, Massachusetts, Arlington, Texas, or Bloomington, Minnesota. The attraction was wide-open spaces with freeway access and lots of parking. Gradually, some cities ceased even to be the nominal sponsors of teams with names like "Minnesota Twins," "Texas Rangers," or "California Angels." For a while, it looked as if suburban options would eventually carry the day and that cities would increasingly lose the ability to use sports teams to instill local pride.

Today, however, downtown locations have become the rage. As massive spaces have become available through urban renewal and deindustrialization, sports arenas have emerged as an important way to "soak up" excess land, often in conjunction with convention centers, hotels, and historic districts. Downtown stadiums first became popular in the 1960s with the construction of multipurpose facilities on formerly industrial "riverfront" locations in Pittsburgh, St. Louis, and Cincinnati. Soon, massive downtown structures appeared in a variety of other cities such as New Orleans, Seattle, and San Antonio. But something was not quite right.

Many of the huge, multipurpose stadiums lacked symbiotic connections to the rest of downtown. During the early years of downtown stadium construction and usage, there was some controversy about just how "downtown" some facilities were. Architecturally, most multipurpose stadiums featured a circular "toilet bowl" design that did not allow fans to have visual contact with the outside world. Some, like the Superdome in New Orleans and the Alamodome in

San Antonio, were enclosed. In addition, most were surrounded by vast parking lots that further separated them from other downtown activities. While they were technically located in the downtown, many fans simply drove in and drove out without spending much time or money at downtown businesses. For a while, promoters in St. Louis even advertised how easy it was to do so.

Of course, some stadiums worked better as downtown catalysts than others. The Kingdome in Seattle, for example, was located just to the south of the historic Pioneer Square area, where several sports bars and restaurants were thriving by the late 1970s. Mariner hats and Seahawk shirts were for sale everywhere. Promoters began to argue that there was money to make by increasing the connections between downtown sports arenas and downtown businesses. Stadium construction was linked to comprehensive urban renewal and revitalization efforts.

Downtown Stadiums and Urban Revitalization

The new generation of downtown arenas took off with the opening of Oriole Park at Camden Yards (Baltimore) in 1992. For a variety of reasons—most of them related to profitability—owners began to demand special-purpose facilities, that is, arenas built for just one sport. Multipurpose stadiums built to house football, they argued, were far too large to attract baseball fans who wanted more intimate surroundings. The new model for baseball (inspired by Forbes Field, an old neighborhood park in Pittsburgh) was an "old-fashioned" place, preferably open at one end for a view of the city skyline. Because baseball games were played so frequently over a long, five-to-six-month season, it was felt that parks could best be located downtown so as to channel fans into nearby cafes and shops. Team owners seeking subsidies for expensive projects pushed the goal of downtown revitalization. Baseball, they argued, would "save" the city. Soon after Camden Yards opened, there were downtown baseball parks in Cleveland, Denver, San Francisco, and Toronto, and similar schemes were on the drawing board for San Diego, Cincinnati, and Pittsburgh.

Professional football got off to a rocky start during the Depression years of the 1930s but took off during the postwar period. Football stadiums are normally much bigger than those built for baseball and may hold twice the number of fans (perhaps 80,000 vs. 40,000). There are also fewer home games and so the day-to-day economic impact on nearby businesses is likely to be less significant. In fact, some have suggested that "football only" stadiums are likely to create a dead zone nearby since games are few and parking requirements are

Coors Field (baseball) in the LoDo section of downtown Denver

great. Still, new and/or redesigned football stadiums are going up in and around many downtowns as well. Cleveland, Charlotte, St. Louis, Indianapolis, and Seattle all have new downtown football stadiums. Some of these are part of convention center complexes or comprehensive sports districts, but a few stand alone.

While some teams are still moving outward to the suburbs, a return to the city is more visible. Football facilities are typically less embedded in the urban fabric than those built for baseball or indoor sports and often occupy relatively isolated corners of downtowns, a wise choice in light of the dead zone allegation.

Venues for basketball and hockey are the easiest to insert into the heart of the downtown, especially since they often use the same arena. Today, approximately two-thirds of the NBA and NHL teams play downtown. Both sports feature lots of home games, thus increasing the potential business for nearby sports bars and theme restaurants. Recently, women's professional basketball teams and indoor soccer leagues have emerged to compete for space on a desirable home court. While other sports may someday seek downtown locations, so far the list is short. Indianapolis has a downtown ballpark for its minor league baseball team, while Columbus has built the only just-for-professional-

soccer stadium in the country in a location near to, but not in, downtown. For now, the famed Boston Garden (rebuilt as the FleetCenter) still epitomizes the quintessential downtown sports venue.

Museums and Cultural Centers

As in the case of sports venues, museums and cultural centers appear to be gradually reconcentrating in downtown locations. Culture sells, and it is playing an ever-increasing role in the marketing of downtown as a prime tourist destination and a desirable place to live. As in the case of sports, a great many early museums and art galleries were not located in the heart of downtown. Many of the most famous museums and galleries in America resulted from the largesse of wealthy entrepreneurs during the late nineteenth and early twentieth centuries—the Carnegies, Mellons, Rockefellers, and others. The establishment of cultural centers coincided with both the City Beautiful Movement and the expansion of streetcar systems into suburban neighborhoods. As a result, museums, theaters, and galleries were often located in parklike settings well away from the "grimy" workaday world of downtown.

New York City's Metropolitan Museum of Art was placed in Central Park, while similar parklike locations were found for museums and other cultural facilities in St. Louis, Boston, San Francisco, Chicago, and San Diego. In Cleveland and Pittsburgh, cultural centers were established in university districts several miles from the central business district. In many other cities, major museums were located on grand boulevards either on the edge of downtown or a few miles out. Los Angeles epitomizes this pattern but Atlanta, Phoenix, Detroit, and Philadelphia are also good examples. In a few cities, such as Indianapolis and Fort Worth, art museums were simply put in nearly random, suburban locations. For many decades, art and business were kept separate. The ideal setting for the latter was a serene, garden setting well away from the hustle and bustle of the city. Symphony halls were more likely to be downtown since they were often converted theaters, but in many cases they were also part of rustic, midtown cultural zones.

Libraries and business colleges were more closely linked to the business and legal needs of downtown. They were often located in the CBD core or in nearby civic centers. Universities, at least those of any size or prestige, were very rarely downtown, although a few, such as Brown University in Providence or New

York University, are only a few blocks away. Students, like art, were best kept away from the profane city.

Museums and the Rise of Downtown Tourism

Over the past two decades, culture as entertainment has begun to play a very important role in the image of many American downtowns. Rising levels of affluence and sophistication have meant that ever-greater numbers of people seek interesting "experiences" in top-quality museums, galleries, and theaters when they visit downtowns for business or pleasure. As a result, there has been a proliferation of new cultural facilities, including aquariums, science centers, history museums, huge library and "interactive information" centers, transportation museums, specialized art galleries, industrial museums and tours, ship museums, and children's museums. In addition, many cities have established a variety of architectural tours as well as excursions that focus on local legends and folklore. Beautiful old theaters have been preserved and new performance centers have been built in all shapes and sizes in order to house an increasing variety of live events. "Big league" cities need to have symphony orchestras, opera and ballet companies, and small, avant-garde theater groups as well as sports teams. Large numbers of retail art galleries and museum shops complement most of these activities.

Not all cultural activities are housed in permanent buildings or even semipermanent sites. Cities have always had parades on major holidays such as Christmas, the Fourth of July, and St. Patrick's Day. Lately, these have been joined by a vast number of ethnic festivals, food fairs, wine tastings, First Night (New Year's) events, folk art displays, car shows, marathons, cycling and rowing competitions, and public concerts. While most of these activities typically take place on downtown streets, they reinforce the demand for good civic spaces such as central squares and waterfront parks. Big festivals cry out for grand settings.

Urban universities have also begun to play a role in the provision of downtown cultural offerings. Some, such as Cleveland State University, have been around a long time and have recently been expanded or greatly enlarged, while others, such as Indiana-Purdue at Indianapolis, are relatively new. In addition, a wide variety of community colleges, art schools, law schools, and business and technical schools have arisen in and around downtown. At least some of these offer cultural and sporting events in downtown facilities.

The performing arts center in Denver

Miscellaneous Attractions: Observation Towers, Amusement Parks, and Marinas

Culture and sport dominate the list of downtown special attractions, but there are other types of places to go and things to see. Observation towers and rooftop revolving restaurants offer stunning views of the city and region, often reinforcing the idea that the downtown and its site are more interesting to look at than most suburban settings. There are few if any such towers at regional malls or office parks. While they are far from common, some downtowns now have amusement parks, botanical gardens, and even zoos. Waterfront districts often not only have the usual festival marketplaces but also battleships, submarines, and historic merchant ships berthed conveniently for tours and special events. Rental boats, water taxis, public and private marinas, and harbor excursions complete the list of possible water attractions.

The St. Louis Arch and boats along the waterfront park

Monuments and major historic preservation projects can also qualify as major downtown attractions. Renovated train stations and power plants, state capitols, grand fountains, and unusual features like the Gateway Arch in St. Louis can help give downtowns a special sense of place. Such features are best displayed, of course, in cities that make use of a stunning, or at least unusually pleasant, physical site.

Megastructures and the Sense of City

At first glance, one might think that it would go without saying that the more major attractions in a downtown the better. The problem is that many major attractions take up huge amounts of space and, if located poorly, they can disrupt or even destroy the small-scale spatial linkages that make downtowns interesting and enjoyable. This is especially true of sports stadiums and huge museum complexes. If these facilities are located close to the city center, it could involve closing off streets and widening adjacent highways. On the other hand, if major complexes are located a bit too far from the city center, they may foster the development of competing nodes that diminish the importance of the core. The vast parking lots and garages required by these facilities could further spread out and dilute downtown activity. Most megastructures, almost by

The Tower of the Americas, Alamodome, and convention center on the edge of downtown San Antonio

necessity, have substantial areas of blank walls facing the street. Even when they are well landscaped, large art museums, science centers, and sports arenas can diminish urban life as well as enhance it.

So the question becomes not only how long is the list of major downtown attractions but how well are they located to support each other as well as other types of activities. Ideally, every big attraction should be part of the urban fabric and well connected to smaller streets and alleys. "Big box" attractions should not be located together in a huge superblock zone, yet they should not be so far apart that they fail to contribute to the overall sense of downtown as one place. With this in mind, I evaluate the downtown case studies.

Variable 7. Major Attractions: Sports, Culture, and Entertainment

Baltimore: A (10)

Baltimore has become the quintessential "city as theme park." Oriole Park at Camden Yards is the model for downtown baseball parks not only because of its old-fashioned design but also because of its location next to the Harborplace recreation zone, convention facilities, large hotels, and a variety of pub-

lic transit, including both light and heavy rail. Not far away, but more peripheral to downtown activity, is a new football stadium for the Baltimore Ravens.
There is also a downtown arena, but no major league basketball or hockey
teams currently use it. The harbor itself is now lined with museums, including
the National Aquarium, the Maryland Science Center, the Baltimore History
Museum, the Civil War Museum, and the Public Works Museum. There are
harbor excursions, a maritime museum, rental boats, and an observation deck
on top of the World Trade Center. Several historic attractions (including Babe
Ruth's birthplace) are not far from the harbor. An official cultural district is
now emerging around the Walters Art Museum and Meyerhoff Symphony Hall
in the Mount Vernon District at the Washington Monument just to the north
of the central business district. The restaurants of Little Italy and the food shops
at Lexington Market are also important downtown destinations. Most of the
larger attractions line the harbor and thus do not greatly disrupt the small-scale
street system in the downtown core, although the concentration of large buildings around the ballpark/convention center area is becoming problematic.

Seattle: A (10)

Seattle has combined a stunning physical site with a plethora of major attractions designed to make use of it. The downtown waterfront region is now
home to an aquarium, a maritime museum, a variety of harbor excursions and
ferries, a new art museum, a performing arts center, a gold rush museum, an
"underground" tour in Pioneer Square, and an Omnidome film experience. At
the northern end of downtown, Seattle Center, the setting for the 1962 World's
Fair, houses the Space Needle, the new Experience Music Project, a children's
museum, an NBA basketball (Supersonics) arena, a science museum, and the
monorail station. South of downtown, the multipurpose Kingdome has been
imploded and is being replaced by separate facilities for professional baseball
(Safeco Field, Mariners) and football (Seahawks Stadium). Meanwhile, a new
history and science museum is under construction adjacent to the expanding
convention center.

Outstanding views of the downtown and surrounding region are available
from the Space Needle to the north and the Smith Tower and the 985-foot Columbia Seafirst Center in the south. Pike Place Market, Pioneer Square, and
several waterfront piers full of shops and restaurants attract tourists and locals
alike. Downtown Seattle has gone from being a rather boring and dreary place
in the 1950s to a major international tourist destination. The list of things to do

Museums, boat rides, and restaurants on the Seattle waterfront

and places to go is long and growing rapidly. The main weakness is the lack of a consensus civic space for large festivals and special events, although the new Westlake Park may help. Otherwise, the pieces fit together pretty well with no place—other than the "two stadium" zone in the industrial area south of downtown—overwhelmed by big-box attractions.

Denver: A (10)

Downtown Denver hosts four professional sports teams, all with "local color" names—the Rockies, Nuggets, Avalanche, and Broncos. Denver has an old-fashioned baseball park (Coors Field) that has acted as a catalyst for helping to revitalize the area around it (LoDo). At the other end of the historic district is the Pepsi Center, a new basketball and hockey arena. Across the Platte River, above the downtown, a modern football stadium is replacing Mile High Stadium (1948), but it is too far away (and on the other side of the freeway) to have any pedestrian impact. Elitch Gardens, a classic amusement park, has relocated to a "next to downtown" site on the Platte River floodplain. Nearby, a children's museum and Colorado's Ocean Journey, a new aquarium, add to the list of attractions. These are clearly peripheral to the downtown core, but plans

are afoot to expand the central business district into the area. At the other end of the downtown, a new, architecturally intriguing art museum sits at the edge of the state capitol grounds, along with new facilities for the history museum and library. Not far away, a massive, four-theater performing arts center abuts the convention center. Across Cherry Creek, a renovated brewery acts as a theme building and social center for the (relatively) new multicampus Auraria Educational Complex. There are some problems linking everything together, but the bus mall connects the capitol with LoDo and light rail helps as well.

Cleveland: A− (9)

Cleveland, once maligned for having a burning river lined with slag heaps, now has a vibrant downtown with a wide variety of major attractions. The most famous of these is perhaps the I. M. Pei–designed Rock and Roll Hall of Fame, which now shares space on a waterfront park with the Great Lakes Science Center. Nearby is a new football stadium for the latest incarnation of the Cleveland Browns. A few blocks away, and much more integrated into the urban fabric, is Jacobs Field, an old-fashioned baseball park opened in 1994, which has spawned a number of nearby sports bars and cafes. Next door is Gund Arena, an all-purpose sport and entertainment complex that is home to the Cleveland Cavaliers (basketball), the WNBA Rockers, and the Lumberjacks (minor league hockey). The waterfront is also home to a restored 1925 Great Lakes freighter that now serves as a museum along with a World War II submarine. Playhouse Square, a group of four restored movie and vaudeville theaters on Euclid Avenue, provides venues for the performing arts. The theaters were slated for demolition until 1970, but now it is one of the biggest renovation projects of its kind in the United States and a real economic boost for the city. Cleveland State University anchors the east side of downtown and its sixteen thousand students have access to the new, first-class downtown library.

A variety of river and lake excursions are available in the Flats entertainment district, and there is plenty of "industryscape" still remaining along the river to give the downtown a unique sense of grungy authenticity. Most of these activities are located within walking distance, but when the weather gets cold, a waterfront trolley and an inside passageway from the train and light rail station to the ballpark and arena help.

The primary negative dimension of the downtown Cleveland cultural scene is that its world-famous art museum and symphony are located at University Circle (Case Western Reserve campus) several miles to the east. Several smaller

Cleveland's renovated Playhouse Square theaters

museums are also there. The connection, through some very dreary neighborhoods, is not great. Still, there is a lot to do downtown.

St. Louis: A− (9)

Downtown St. Louis has a long list of major attractions, but it sometimes
seems that the sum of the parts is greater than the whole. Three of the major
sports are represented with NHL hockey (Blues), major league baseball (Car-
dinals), and professional football (Rams), each occupying its own facility.
Busch Stadium (baseball) is a 1960s multipurpose venue, but the Kiel Arena
and the TransWorld Dome are relatively new. The latter was built in association
with the city's massive new convention center. While all of the sports complexes
are centrally located and reasonably well connected to mass transit and supporting activity zones, they are far enough apart to have separate spheres of influence. Both the TransWorld Dome and Busch Stadium are located near the
Gateway Arch (with its park and museums) and Laclede's Landing entertainment district, while the Kiel Arena is close to St. Louis Station, with its many
cafes and shops. Still, the spaces between these major attractions usually seem
to be underutilized, and people usually head for nearby large parking garages
after a game.

The Kiel Arena on a light rail line in St. Louis

The Union Station complex, once the largest train station in America and now the largest single historic preservation project, is a major attraction in its own right and draws millions of visitors each year for food, shopping, movies, and music. Connections are tenuous, however, and relatively few people stroll from the station further into downtown. Along the riverfront, riverboat gambling casinos, supported by various boat rides and excursions, attract eager crowds of tourists and locals. As in Cleveland and Pittsburgh, however, "culture" is not downtown, since the major art galleries, museums, and theater and symphony venues are located either in Forest Park or the West End several miles away.

Indianapolis: A− (9)

Not long ago, there was very little reason to visit downtown Indianapolis. Today, however, the city ranks fairly high on the list, with several major attractions. The NBA Pacers occupy the just-completed Conseco Field House downtown, while the NFL Colts are housed in the (60,000-capacity) RCA Dome, built in association with the convention center in 1983. Downtown sports attractions also include minor league baseball (Victory Field) and a huge amateur sports complex (for swimming, tennis, etc.) at the relatively new (1969)

campus of Indiana-Purdue Universities at Indianapolis. The Indianapolis Zoo has recently moved downtown to the revitalized White River waterfront. Nearby, a museum of Native American art, a history museum, and an Imax Theater provide destinations along a system of canals with pedestrian promenades and boat rentals. The symphony occupies a renovated downtown theater, but the main art museum is several miles to the north. Additional attractions are a (re)renovated train station/festival marketplace (Union Station), City Market, Monument Circle, and the state capitol.

Atlanta: A− (9)

Downtown Atlanta has a lot to offer, but the attractions are widely scattered and do not always contribute to a coherent sense of downtown as a place. The 1996 Olympics played a major role in rebuilding the city's sports (and other) infrastructure. The Georgia Dome, a megastructure that helps to delimit the western edge of downtown, houses the NFL Falcons, while a new "built for the Olympics" baseball stadium (Turner Field) sits well to the south of the downtown core. The new (1999) Philips Arena is home to the NBA Hawks and NHL Thrashers and is connected to the massive CNN Complex. CNN offers studio tours of its news broadcasting operation as well as shops and a food court. Downtown Atlanta is also ringed by universities, some of which provide cultural and sports attractions. Georgia State occupies a campus made up largely of old office buildings in the center of town, while Georgia Tech, Emory, Spelman, and Morehouse are nearby. Underground Atlanta and the World of Coca-Cola museum next door bring visitors into the core of downtown. Most of the cultural attractions are located to the north in Midtown (sort of an extension of downtown), including the Robert Woodruff Arts Center (comprising the High Museum of Art, Alliance Theater, Symphony Hall, etc.) and the botanical gardens in Piedmont Park. The main history museum is even further north in Buckhead. Just to the east of downtown, the Martin Luther King Jr. National Historic Site is a major attraction dedicated to the struggle for civil rights. The Science and Technology Museum (Scitrek) is also a major downtown attraction, along with the state capitol.

Phoenix: A− (9)

The question in Phoenix is "how much territory can a downtown occupy?"—especially when there are vast empty spaces between major nodes. Downtown Phoenix has quite a few major attractions, but some, like the state

Bank One Ballpark, America West Arena, and Patriot Square in downtown Phoenix

capitol, the art museum, the Heard Museum (native cultures and art), and the Mining and Mineral Museum are located well away from the core on long, wide highways. On the other hand, the central part of downtown includes the huge new BOB (Bank One Ballpark, 1998) for the Arizona Diamondbacks and the America West Arena for the NBA Suns. So far, however, relatively little in the way of a support district has grown up around these attractions, and so symbiotic relationships are not always achieved. Nearby, the Arizona Science Center, Heritage Park, the Museo Chicano, and Symphony Hall add to the list of core area attractions that ring the civic plaza. A problem arises, however, with the fact that two arenas and a convention center in close proximity make for a "big box" downtown with few cozy nooks and crannies.

Pittsburgh: B+ (8)

There is little room in the compact core of downtown Pittsburgh for space-extensive facilities such as sports complexes. As a result, the new (Steelers) football stadium and (Pirates) baseball park are being built across the Allegheny River, where the multipurpose Three Rivers Stadium once stood. To the east, an arena for the NHL Penguins lies uphill from downtown in a 1960s urban re-

newal district. While all of these facilities are close to and visually connected to the downtown core, it can be a long cold walk across the bridge before or after a game. The city plans to expand the downtown across the river with a grid-pattern district and new mixed-use buildings. Some progress has already been made, and the Andy Warhol Museum and the Pittsburgh Science Center are located there. Two splendid old theaters, including Heinz Hall (1926), have been renovated in the downtown core for the symphony and other cultural events. As in Cleveland, however, the major art gallery and most of the museums are located in a "university/cultural district" (University of Pittsburgh and Carnegie-Mellon) in the Oakland neighborhood several miles to the east. The old industrial buildings in the Strip have been discovered as possible sites for museums, however, and the Pittsburgh Regional History Museum currently occupies one of them. Duquesne University occupies a hilly site just outside the downtown core, but pedestrian connections are not good. Station Square, a renovated train station/festival marketplace, is located just south of the downtown core, across the Monongahela River, and is the starting point for several types of harbor excursions. Nearby, the Monongahela Incline (railway) leads to a series of observation points as well as some nice "city view" restaurants.

Minneapolis: B+ (8)

Minneapolis shares its metropolitan area with its twin city, St. Paul. Similarly, the two downtowns share functions, with Minneapolis being the business center while St. Paul is the political capital. Nevertheless, downtown Minneapolis is a vibrant place. The NFL Vikings and the nearby University of Minnesota Golden Gophers play in the Hubert Humphrey Metrodome, as does the Minnesota Twins baseball team (at least for now). The NBA Timberwolves play in the Target Center a few blocks away. The two major art museums are located on the edge of downtown, with the Walker Art Center (which includes the Guthrie Theater) and Sculpture Garden near Loring Park being the most accessible. The downtown library houses a planetarium, but there are few other major downtown museums. Several theaters, such as the State and Orpheum, have been renovated for cultural events and Orchestra Hall is also located downtown. The massive (47,000) University of Minnesota is not far from the downtown core, but the connections are not great. The rapids and locks along the Mississippi River make for an interesting physical and cultural attraction, but it is not really a big draw.

Sculpture garden at the Walker Art Museum on the edge of downtown Minneapolis

San Antonio: B (7)

There is no doubt that downtown San Antonio is a major attraction for tourists and locals alike, but it is not easy to come up with an extensive list of individual major attractions. The whole is simply a lot greater than the sum of its parts. The Riverwalk with its assortment of restaurants and music venues is *the* major attraction, but there are supporting players. The Alamo is the major historic icon in the city if not the entire state of Texas. Hemisfair Plaza, the erstwhile world's fairgrounds (1968), contains a collection of shops and museums, including the Institute of Texan Cultures, as well as the Tower of the Americas, a six-hundred-foot opportunity for a view of the metropolitan area. The Alamodome (1993) stands at the edge of downtown, but the Spurs (NBA) have moved out and there are no other professional sports teams to use it. Special events will no doubt take up some of the slack. Also on the edge of downtown, there is a new library and an interesting art museum that is housed in a renovated brewery. To the east of the city core are the (Mexican) market district and the new downtown campus of the University of Texas at San Antonio. San Antonio barely has a major league downtown, but it is full of visitors. The attraction is the ambience.

Portland: B (7)

Portland has a very attractive and much-visited downtown, but the list of major special attractions is not particularly long. The only professional sports venue is the Rose Garden (NBA Trailblazers) and it is located across the Willamette River in a subdowntown district. The Oregon Museum of Science and Industry is also across the river in a relatively remote, semidowntown location. Cultural attractions in the core include the Oregon History Center, the Portland Art Museum, the Police Museum, and the Oregon Maritime Center and Museum. Special events take place in the civic auditorium and in new and renovated theaters in a performing arts complex. River and harbor excursions are available along the waterfront and marina, and the Ira Keller Fountain is a unique attraction, as is the Portland Building. Portland State University (17,000 students) is located near the center of downtown and contributes to the life of the city. On the west side of the river, at least, major attractions are exceptionally well connected by a cozy grid of lively streets.

The Ira Keller Fountain in downtown Portland

San Diego: B— (6)

Downtown San Diego lies just to the south of the two-square-mile amenity known as Balboa Park. In evaluating the downtown's major attractions, much depends upon whether or not the park is included in the definition of down-town, since several museums and theaters are located there along with the world-famous zoo. Indeed, the park itself ranks as a major attraction. Because Balboa Park is so large (the museums are at least a mile and a half from city hall), with few routes in from downtown, and because it is separated by a busy freeway interchange, I chose not to include it. (Otherwise San Diego would score an 8 or 9.) I had to draw the line someplace. Given that decision, down-town San Diego does not rank very high in major attractions. There are no sports venues, although a downtown baseball park is scheduled to open in 2004. Plans for a new library have so far gone nowhere.

The primary downtown museum is the Maritime Museum, which consists of three ships, including the 1863 tall ship *Star of India*. There is also a small children's museum, which occupies an old warehouse, and it is possible to tour visiting navy ships from time to time. The waterfront itself, however, is a ma-jor attraction with its tuna boats, luxury cruise liners, and aircraft carriers. There are also a wide variety of harbor excursions and water taxis. Seaport Vil-lage, with its waterfront promenade and cute shops, attracts crowds all year. The Civic Center auditorium houses the opera and other performances, and the symphony occupies its own facilities (although it nearly went bankrupt) in a renovated theater. San Diego City Community College occupies a corner of downtown.

Columbus: B— (6)

As the new millennium began, Columbus was in the process of building not only a new arena for its expansion NHL team, the Bluejackets, but an entire "arena district" to go with it. Still, it is just barely a major league downtown. Across the Scioto River, a massive new Center of Science and Industry (COSI) functions as a major attraction. There is also a very pleasant, if not huge, art museum in the downtown core. There are five major theaters downtown, in-cluding the magnificently renovated (1928) Ohio Theater, used for ballet and symphony performances. The state capitol has recently been restored to its mid-nineteenth-century splendor and attracts many visitors. A recently ex-

panded library is available for use by downtown students from Franklin University, the Columbus School for Art and Design, and Columbus State Community College. Fun-loving visitors, however, tend to concentrate in the Brewery District/German Village or the Short North district at the edges of downtown, and this will be even more the case when the north-side Arena District is completed.

Charlotte: C+ (5)

North Carolina has long been a state with lots of medium-sized cities but no one place with big city lights and major attractions. Charlotte would like to change that. In recent years, the city has emerged as a metropolitan area with the threshold population and the enthusiasm to support at least a few major attractions, and some of those are downtown. Ericsson Stadium (75,000) is the home of the NFL Panthers and occupies an entire corner of downtown, but a moderately large dead zone separates it from the core. Professional basketball and hockey, however, are played in a suburban arena. Other significant downtown attractions include the Mint Museum of Art, Discovery Place, with its Omnimax Theater, and the Museum of the New South. There is also a recently expanded downtown library. The core attractions are close together and easily accessible, but there is relatively little in the way of interesting urban life in between. Charlotte is a new city that is only now gaining a sense of urbanity.

Providence: C (4)

Downtown Providence is tough to evaluate fairly because it is so much smaller than the other districts being examined. If an area similar in size to that covered by several of the other downtowns were examined in Providence, it would include Brown University, with its several museums, galleries, and historic buildings, as well as some nearby residential neighborhoods. It seems only fair, therefore, to at least mention these nearby attractions in the downtown category. There is simply not much room in the core for space-extensive major attractions. The state capitol high on a hill overlooking downtown is a major focal point, and Water Place Park is becoming a pleasant gathering place. The Rhode Island School of Design has an art museum, and the Providence Art Club is next door. Minor league hockey is offered at the civic center, and there is a small, renovated theater for cultural events. The campus of Johnson and Wales (culinary) University occupies a new campus on the site of an old downtown department store, but in general downtown lacks "big draw" centers.

Summary: Major Attractions

Baltimore 10
Seattle 10
Denver 10
Cleveland 9
St. Louis 9
Indianapolis 9
Atlanta 9
Phoenix 9
Pittsburgh 8
Minneapolis 8
San Antonio 7
Portland 7
San Diego 6
Columbus 6
Charlotte 5
Providence 4

Historic Landmarks, Districts, and the Sense of Place

The creation of a historic sense of place has become a very important component of any definition of a revitalized American downtown. Led by such cities as Boston, New Orleans, and San Francisco, downtown planners now focus nearly as much attention on the creation of a human-scale, historic, entertainment district as they do on the skyscrapers of the financial core. History and place have become very saleable commodities. In addition, the smaller, older buildings normally found in "olde towne" can contribute to a cozy, pedestrian-friendly environment that provides a distinct contrast to the sterile towers of the central business district. Initially, at least, the cheaper rents found in semi-renovated buildings on the edge of downtown can facilitate the formation of a district full of art galleries, cafes, jazz joints, and used book stores. Even if too much success kills these avant-garde businesses, the character of the area and its cutting-edge reputation is likely to remain for a long time. Greenwich Village, the French Quarter, and North Beach are still known as *places*, even though they no longer have much of their original character.

Downtowns have very different historic potentials regardless of their his-

toric preservation policies and programs. Some are simply a lot older than others and thus have a much greater stock of old buildings, small streets and blocks, and real and/or legendary "history." For example, in 1900, Baltimore and St. Louis had over 500,000 people, while most of the other cities on the list had between 100,000 and 200,000. San Diego, on the other hand, had only 20,000, while Charlotte (18,000) and Phoenix (5,000) were even smaller. Of course, age does not guarantee a good stock of historic structures, since decades of urban renewal and highway clearance projects are likely to have taken a toll. Nevertheless, it is difficult to build a historic district from scratch even if authenticity is not a high priority. Most good downtowns, however, have some sort of historic dimension if not a full-fledged historic district.

City Structure and the Zone of Transition

Most of the classic models and generalizations about the spatial structure of North American downtowns have included some recognition of a marginalized "zone in transition" adjacent to the core of the central business district. The disheveled appearance of the transition zone was due to the expectation of change. The downtown core, with its skyscrapers and big department stores, loomed large over the small and usually underbuilt and undermaintained buildings nearby. Because of the popularity of the skyscraper, American downtowns were typically overzoned, that is, zoned for much greater densities than were actually prevalent in most of the district. Consequently, the owner of a two- or four-story building on the margins of downtown would simply try to hold on until demand built up for a new skyscraper and then sell the building, and more importantly the lot, for a tremendous profit. But since there were so many buildings and lots, the likelihood of such a sale was usually slim, and so most older buildings sat for years if not decades "waiting their turn." In the meantime, owners would simply "milk" the structures by collecting whatever rents could be had while investing nothing in upkeep, since, after all, they would soon be torn down.

The greedy behavior of building owners, however, was only part of the problem. After the many catastrophic fires of the nineteenth century and the San Francisco earthquake and fire of 1906, American cities developed strict building codes. The idea was to encourage the replacement of older structures with newer and better ones that were fire and earthquake proof. Buildings that were not up to code were "redlined" by banks, insurance companies, and other financial organizations. Even owners who wanted to maintain buildings could

not get financing to do so. Buildings that were not safe according to sometimes arbitrary codes were not eligible for insurance, and so most businesses were leery of them. With few interested buyers and with mortgage loans unavailable, many owners had no choice but to hold on and wait.

By the 1920s, however, the combination of skyscraper construction in the core and suburban development at the urban fringe was taking demand away from many parts of downtown. A few towers and large department stores served to concentrate downtown activity on just a few blocks, while space-extensive activities increasingly moved out of downtown altogether. The wait for change in the zone in transition became much longer. The downtown fringe became synonymous with "slum" by the 1950s, and many such areas were slated for urban renewal.

Historic Preservation and a Sense of Place: Changing Aesthetic Priorities

By the mid-1960s, most American downtowns seemed all too eager to lose their identities. The combination of modernist, "glass box" office towers, "slum" clearance under urban renewal, and highway construction projects erased much of the historic identity that was once writ large in central cities. City leaders took pride in their abilities to achieve progress, and progress usually meant getting rid of the baggage of the past. To a very real degree, the enthusiasm for historic preservation that arose in the mid-1960s was a direct reaction to the placelessness of the new downtowns.

Buoyed by the success of preservation efforts in such cities as Charleston, South Carolina, and New Orleans and the popularity of a few "role model" projects such as Ghirardelli Square in San Francisco, an increasing number of civic leaders and community groups began to argue for a new aesthetic. Preservationists lobbied for new policies and procedures that were more sensitive to the maintenance of historic identity. With the 1976 Bicentennial on the horizon, history became something to recognize and celebrate rather than something to obliterate and pave over. In 1966, the U.S. Congress passed the National Historic Preservation Act, which went a long way toward changing the rules. The act made it tougher for government agencies to destroy buildings that were listed on the National Register of Historic Places, and various tax breaks were developed to make preservation projects feasible. In addition, so many older buildings had been torn down in the previous decades that by the late 1960s those that survived were beginning to have scarcity value. Instead of a sea of

decay, civic leaders saw precious remnants. Some of these remnants became important landmarks, while others anchored incipient historic districts.

What to Preserve: Landmarks, Skid Row, Industrial Buildings, and Houses

Often the development of a new preservation ethic in a particular city revolved around the saving of one or more significant downtown landmarks. Citizens rallied around the cause as beloved theaters, churches, department stores, train stations, hotels, or even city halls were faced with possible destruction. Preservationists scurried to make sure threatened structures were listed by the National Trust and that owners were aware of possible tax advantages, loans, or zoning variations that could be used to keep older buildings economically viable. Most downtowns have at least one major success story in the form of a beautiful and viable rehabilitation and reuse—perhaps an old theater now used by the symphony or a former train station that now functions as a festival market. Sometimes newly formed historic site boards played a role in delaying demolition until a new buyer or new financing could be found. By the mid-1970s, it was becoming clear that old did not necessarily mean obsolete and that a sense of history could even be used in promotional literature to attract tourists and new businesses.

The next highest priority after the saving of major landmarks was often the rehabilitation of the old downtown core, an area that had become a zone of discard or even a skid row. Saving these areas was typically more controversial, since the remaining architectural charm was embedded in not only the usual patina of age but also garish signage, ugly graffiti, and sometimes a large and scruffy population. It was not like saving the train station or the city hall; this was a vague district with a poor reputation and a high crime rate. The population itself tended to be controversial as well, since everyone either loved or hated the idea of expelling local residents in order to save buildings. During the halcyon days of urban renewal, residents of downtown hotels could be tossed out with little effort, but by the 1970s, new, more sensitive (and sensible) regulations required relocating people and maintaining social services. The controversy continues. Some downtowns have made an effort to rehabilitate low-income single-room-occupancy hotels and even build new ones in order to keep at least some options open for displaced residents. Others have not.

By the time the Bicentennial rolled around in 1976, it was clear that "fun" cities had historic districts. The French Quarter was an excellent role model for

the development of historic entertainment districts, but for a long time it was hard to dispel the notion that New Orleans was somehow exotic and different and therefore not easy to emulate. San Francisco's Chinatown and North Beach were also famous and lively districts with lots of historic character, but that city too was perceived as unusual. Gradually, many cities learned that emulation was possible. During the 1970s, Pioneer Square in Seattle, the original "skid row," was discovered and gradually fixed up. Old Sacramento became a state park, and city leaders began to talk of a Gaslamp Quarter in San Diego. From Larimer Square in Denver to Underground Atlanta, once disheveled and nearly forgotten downtown commercial districts were turned into entertainment zones. By the early 1990s, some wags suggested that all you had to do in any American city was to simply hail a cab and say "take me to old(e) town(e)."

Not only did former zones of discard usually have a variety of older buildings with rents low enough to attract artists, cafes, and jazz clubs, they also had reputations that reinforced the prospect of finding wild and crazy fun and entertainment. For a while, the old skid rows, with their former bordellos and speakeasies, were seen as just slightly scary, yet still attractive to a growing contingent of urban fun-zone pioneers. As the districts grew and gained fame,

The Mercado district in San Antonio

however, they came to be seen as ideal tourist destinations. They also became safe with beefed-up police and private security: skid row as theme park.

Occasionally the issue of authenticity arose but usually only in passing. The early districts often emphasized not only historic buildings but historic establishments and activities—such as Preservation Hall in the French Quarter. As time passed, however, a variety of chain establishments, sometimes specializing in locations in old buildings in former skid rows, began to appear—the Old Spaghetti Factory, Rock Bottom Brewery, Hooters, Hard Rock Café, etc. As historic districts became more predictable, they came to be valued as much for their human scale and walkable ambience as for their excitement and history. Today, they are often seen as the perfect alternative to the sterile, glass-and-steel towers of nearby financial districts. Sidewalk cafes, galleries, small theaters, and unusual gift shops provide a cozy urbanity heretofore missing in most American downtowns.

The trend in recent years has been to enhance lively, historic entertainment districts with major attractions such as sports arenas and science centers. The best examples might be Coors Field in Denver's LoDo district or the new sports complexes just to the south of Seattle's Pioneer Square. The goal, of course, is to create a multifaceted set of downtown attractions that will bring in big-spending tourists and locals alike. As discussed above, however, as major attractions become more self-contained, such symbiotic relationships become increasingly problematic.

Zealous preservation efforts have also brought about the rehabilitation and reuse of industrial buildings, warehouses, and even port facilities in some downtowns. Former breweries, grain elevators, meat packing plants, factories, and shipping piers have all been used to create and sell local history and a sense of place on the once marginal downtown fringe. Following the example of Ghirardelli Square (1964), Quincy Market, and New York's Fulton Fish Market, former factories and warehouses were rapidly converted to festival markets, and new "old markets," such as those in Baltimore's Harborplace, were constructed as well. Since developer James Rouse was often involved, the process of urban revitalization through reuse was sometimes referred to as "Rousefication"—or "Quincification" after the Boston example. It should be no surprise that successful projects are quickly emulated in American cities whether they be malls or festival markets. Soon old ships, railroad cars, trestles, cranes, and bridges were added to the list of treasures and sometimes incorporated into museums or used for waterfront cafes.

Historic industrial landscapes in the Flats district of Cleveland

Returning to the theme of authenticity for a moment, it is important to mention that some cities have attempted to keep industrial zones functioning as vital support districts and even alternative employment centers for the new downtowns. Lake Erie freighters still sail up the Cuyahoga River just missing the extended porches of nearby nightclubs. Portland has zoned a large district as a protected industrial refuge. Downtowns still need computer repair services, machine shops, garages and mechanics, storage units, and the like. It doesn't make sense for these activities to be pushed out to the suburbs.

Historic districts have also played an important role in the zone of assimilation, although they are usually low profile compared to the types of places discussed above. In most cases, it is rare to have an older, intact residential neighborhood within the area that can reasonably be defined as downtown. More often, only a scattering of houses remains, usually with apartments and commercial activities interspersed. If the houses and apartments have been maintained or upgraded, the neighborhood may be seen as a quiet residential area rather than a downtown fun zone, but hybridization sometimes occurs. Coffee bars, gift shops, art museums, and galleries often coexist with historic houses in these "midtown" districts. Perhaps the most famous examples are

Boats, ships, and historic industrial landscapes in Portland

Beacon Hill in Boston, Nob Hill in San Francisco, Rittenhouse Square in Philadelphia, Yorkville in Toronto, or Grammercy Park in New York City. These neighborhoods have historic charm, but they are harder to emulate than those modeled after downtown entertainment districts. Very often developers borrow their "history" and charm in order to construct desirable new downtown residential buildings such as in Seattle's Belltown, San Diego's Uptown, or Atlanta's Piedmont Park area.

Signage, Statues, Events, and Ambience

Some cities exude historic character even in the relative absence of extensive historic districts or famous historic landmarks. Sometimes scale is the key. Cities with small blocks and narrow, winding streets lined with moderate-sized buildings seem more historic and "placeful" than those with wide and busy streets and skyscrapers. Occasionally cities succeed at capturing the essence of a regional identity through the use of color, vegetation, and modern variations of a particular architectural style. Public art in the form of statues, murals, fountains, and monuments can also help to connect residents and visitors to a sense of time and place, especially when they celebrate local personalities or traditions. "Mexican" markets, "German" beer gardens, "Chinatowns," and

"Little Italys" can also enhance a sense of local culture. Signs, smells, music, architectural embellishments, and special events can all be used to play up these local place characteristics.

Historic preservation nearly always involves a trade-off between the display of a particular era frozen in time, on the one hand, and the representation of the effects of the gradual passage of time on the other. An example of the former is Old Sacramento, a state park presenting the look of the city during the California gold rush. An example of the latter is the area fronting the Alamo in downtown San Antonio, a diverse yet vaguely historic district featuring buildings and signage from many eras. Historic districts, murals, statues, special events, and ethnic signage almost always raise the question of whose past is being displayed, but this is probably a more important question for academics than for typical visitors.

The irony is that the older the city, the more potential "history" there is to display. But in America, older cities are the ones most likely to have been decimated by urban renewal, ghettoization, and abandonment. Young cities, on the other hand, are more likely to have newer, smaller, and less interesting landscapes, but they have not experienced the same degree of decay. Cities can only save what they have, and some simply have a lot more to work with than others. On the

A mix of new and old architecture in downtown Minneapolis

other hand, what is historic? Gas stations, neon, motels, drive-ins, factories, churches, and tenements may all be considered to be historic by some admirers. Defining a sense of place can be even more difficult. Very few American downtowns have anything like the historic character we associate with European cities such as Venice or Paris, and so it is necessary to have a different standard for success on this side of the Atlantic. Still, many cities are doing pretty well with what little history they have left. Below is my attempt to evaluate these efforts.

Variable 8. Historic Landmarks, Preserved Districts, and a Sense of Place

Seattle: A (10)

Seattle gets high marks for historic preservation, in part because it has managed to combine commercial success with a high degree of social equity. Pioneer Square, the original "skid row," has become one of the best downtown historic districts in the country due to a variety of enlightened policies over the past thirty years. Seattle "pioneered" the use of flexible code enforcement, minimum maintenance requirements, highway down zoning, public revolving funds, and urban beautification projects to revitalize a notoriously sleazy district. At the same time, old hotels were rehabilitated for low-income housing, and a variety of social services such as community health stations were established to help existing residents. As a result, Pioneer Square has become economically viable while maintaining much of its character. The same is true of Pike Place Market. Slated for destruction in the late 1960s, city residents signed petitions to keep the massive 1910-era market going. As revitalization led to success, rules were established to keep the traditional produce activities dominant and limit the number of tourist shops. Pike Place Market is now a major tourist attraction and the regulations have been only partly successful. Still, a lot of local character remains.

Along the waterfront, piers have been recycled for a variety of uses, from restaurants to an aquarium, and the traditional ferry landing has been joined by a number of harbor excursions. To the southwest of Pioneer Square, "real" port facilities still operate, and so the waterfront, while pleasant and accessible, has not been entirely deindustrialized; ships, grain elevators, and cranes add interest to the edges of downtown. In the CBD, a variety of older structures have been revitalized, including the L. C. Smith Tower, one of the nation's earliest (1914) skyscrapers.

Seattle's Pioneer Square Historic District with sports bars for the ballpark and stadium to the south

Cleveland: A− (9)

Downtown Cleveland not only has most of the usual features associated with urban preservation but it has also preserved some of the best downtown "industryscapes" in America. Now that the smoke has cleared and the infamous river and lake pollution has been cleaned up, the massive bridges, cranes, and wharves that make up the Port of Cleveland give the city identity and historic interest. Most of these features are located in a historic area known as the Flats, a low-lying district where the Cuyahoga River enters Lake Erie. The old factories have been converted to restaurants, comedy clubs, and condos, but the ambience of heavy industry remains. Giant Great Lakes freighters still use the river and add visual excitement to the place. Loft housing in the nearby Warehouse District reinforces the historic industrial character of the west side of downtown. On the other end of the CBD, Playhouse Square, a group of renovated theaters now used for live productions, displays the cultural side of the city. The core of the downtown now includes a number of older buildings with sports bars and cafes associated with nearby sports venues. The Terminal Tower (1927), one of America's tallest early skyscrapers, dominates the entire down-

town. Still, much of the eastern edge of downtown has been decimated by decades of decay and urban renewal.

Pittsburgh: A− (9)

Unlike Cleveland, Pittsburgh has removed much of its downtown industrial heritage. As the first American city to use urban renewal to change its image, a concerted effort was made during the 1950s and 1960s to get rid of downtown's smoky, industrial appearance. The Point became a green park and the Hill District was partially cleared for a civic arena. Still, many interesting historic buildings remain. The northern section of the downtown core, occupying a separate grid, has been designated the Cultural District and includes a variety of attractive commercial buildings from the turn of the twentieth century. The Strip, an area of warehouses east of the core, still houses produce markets and taverns as well as a growing number of restaurants. Station Square, already discussed above as a major attraction, is certainly an interesting bit of history as well, especially when combined with the nearby funiculars. Barges still ply the rivers, and nearby neighborhoods are gradually being discovered as sites for recycled brewpubs and cafes. In the CBD core, an excellent stock of older civic and commercial buildings gives the city lots of character, although plans are afoot to destroy a good many of them. So far, the "new" Pittsburgh seems to have chosen to downplay its past, but there is still plenty of history left.

Baltimore: A− (9)

Baltimore is an old city with lots of brick row houses and old commercial districts. Its main attraction, the waterfront or Inner Harbor, has experienced massive clearance close to the downtown core, but it is now being enhanced by preservation planning in peripheral districts such as Fells Point. The core of the CBD was largely cleared by urban renewal for the construction of the modernistic (but now dated) Charles Center, but there are still quite a few small, older structures. The northern section of Charles Street, dubbed the Mount Vernon District, has fared better and includes an excellent stock of historic structures focused on the Washington Monument. Lexington Market, like other markets in central Baltimore, helps to keep a strong sense of historic activity alive. Little Italy, just to the east of the downtown core, displays not only an ethnic identity but has utilized a large stock of existing residential and commercial structures. Many of the museums in and around the downtown core celebrate the city's heritage, as do the historic sailing vessels on the waterfront.

Older buildings in Pittsburgh's Cultural District

Sometimes a sense of history remains even when the actual buildings have been removed, as on Federal Hill Park, the site of an old fort.

San Diego: A− (9)

San Diego is a relatively young city, but a sizeable hunk of its late-nineteenth- and early-twentieth-century core remains intact. With plenty of room for expansion, the small downtown simply migrated away from the waterfront during the early 1900s rather than rebuilding on site. The city's conservative leadership avoided using grand urban renewal projects during the 1950s and 1960s, and as suburbs boomed, the historic downtown was more often forgotten than remodeled. When the preservation ethic arrived in the 1970s, a good selection of older structures still remained. An eight-block section of what was once the main drag, Fifth Avenue, now constitutes the core of the historic

A street festival in San Diego's historic Gaslamp district

Gaslamp Quarter, a successful area for nightlife, restaurants, galleries, and specialty shops. The Gaslamp was slow to take off, but with the opening of the nearby Horton Plaza shopping center in 1985, the district boomed.

Preservation activity has recently spread to adjacent streets, and lofts and galleries are now common. However, some buildings have been destroyed for the baseball park near the Gaslamp Quarter, and empty fields remain too dominant. At the northern end of downtown, Little Italy is experiencing a combination of preservation and new construction as another nightlife and restaurant district. Tuna fleets and navy ships add authentic activity to the waterfront, along with the three ancient vessels that make up the maritime museum. Along Broadway, the main street since the 1920s, several preservation projects have succeeded in making old landmarks economically viable. The U.S. Grant Hotel and the Spreckles Building (including a theater) are prime examples.

Portland: A− (9)

Starting in the 1940s, Portland destroyed much of its historic waterfront to construct a highway. Now that a waterfront park has replaced the highway, the city is much nicer, but the historic link to the river has been diminished. To the north of the CBD core, however, the Pearl (warehouse) District is being revital-

ized with a combination of restored lofts and new apartment construction. Union Station, although a bit off-center, is also being restored. Also to the north, the Old Town/China Town area is largely included in a historic district. Next to the CBD core, the Yamhill Historic District houses a variety of cafes and shops for nearby office workers. Portland has done an excellent job of recycling early-twentieth-century office buildings, and the core seems much more diverse and textured than those of most other cities its size. The old courthouse acts as a focal point for the retail district. Although the southern edge of downtown experienced massive urban renewal in the 1960s, there are some nicely restored buildings on the downtown campus of Portland State University. There is a good stock of still-used industrial buildings across the river, and grain elevators and ships add a sense of functional authenticity to the northern waterfront.

Providence: A− (9)

Downtown Providence, with its narrow, winding streets and excellent supply of nineteenth-century commercial buildings, such as the Weybosset Arcade (1828) and churches, generally exudes a historic ambience. While there is no focused historic entertainment zone, à la LoDo or Pioneer Square, the restored structures along the "mile of history" on Benefit Street and the nearby Brown University area reinforce the downtown's connection to the past. In addition, many of the newer structures in and around the core include architectural references to the city's red brick American Baroque ambience. While some of the historic buildings are underutilized and only partly restored, there is little sense of large-scale abandonment. The potential is there for a more concerted preservation effort. However, the downtown riverfront has been completely redesigned and there are few reminders of past functions. South of the CBD core, a warehouse district is catching on as a potential historic focus, and to the north, the historic state capitol looms over the CBD from its hilltop location.

San Antonio: B+ (8)

Even though they get nearly the same "score," the downtowns of San Antonio and Providence are very different. Providence has lots of historic architecture but has not yet found a way to use it all, while San Antonio squeezes every drop of history from its landscape in order to (over?) celebrate its past. While the Alamo is certainly an important historic landmark, much of the "historic" landscape has been manufactured, via the creation of "Spanish" architecture that never existed when the region was part of Mexico. Nevertheless, there are

A street scene of older commercial buildings in
Providence

a number of nice old commercial buildings and hotels that give the downtown
a sense of cozy quaintness. Mexican markets and restaurants keep alive a functioning "history," especially when combined with strolling mariachis. A former
brewery has been converted to an interesting art museum, and colorful skyscrapers from the 1920s still dominate the skyline. The King William Historic
District and an emerging area for galleries are just to the south of the CBD but
separated from it by the Hemisfair Plaza.

Denver: B+ (8)

Downtown Denver has become well known for the symbiotic relationship
between the LoDo/Larimer Square Historic District, Coors Field (baseball),
and the renovated train station. Cafes, shops, and loft housing have all boomed
during the last decade and the street life there is active. There is relatively little

historic identity in the CBD core, however, due to massive urban renewal schemes during the 1960s and 1970s and the boom in office building construction. There are, however, a few notable landmarks such as the Brown Palace Hotel, the Tivoli Brewery, and the Daniels and Fisher Tower. While there are some nice older apartment buildings and restored houses east of the state capitol in the Swallow Hill Historic District, most of the preservation efforts have been focused on LoDo. The vast fields of the South Platte River Valley and the sprawling Auraria campus separate the downtown core from many of the city's older neighborhoods.

St. Louis: B (7)

It is tempting to say that until the 1950s St. Louis had way too much "history." The downtown was one of the oldest, largest, and most congested in the United States. (Unlike Baltimore, San Francisco, and Boston, it had not been redesigned by fire.) It was certainly the most intact historic core in this study. Time, urban renewal, and disinvestment, however, have taken a toll. Much of the downtown core has been cleared for street widening, civic structures, sports arenas, highways, and the Gateway Arch. Much of what remained after the heyday of renewal was left to decay. Laclede's Landing, a historic remnant of the waterfront warehouse district, was the focus of most downtown preservation activity for many years. In spite of all the clearance, however, there is still a tremendous stock of interesting old buildings ripe for preservation. Although many buildings are still derelict, preservation efforts have accomplished a great deal in recent years. A loft district has been created along Washington Street and massive warehouses are being renovated on the southern edge of the CBD core.

The old post office and other civic buildings are also being revitalized. Neighborhood preservation projects, such as the market at Soulard and Lafayette Square, are not very far from downtown but are poorly connected to it. Union Station has to count as one of America's grandest preservation successes, but it is now so rebuilt that it best fits under the major attraction category. Too much clearance, however, makes the CBD seem like it is located in a vast field, remote from the rest of the city.

Minneapolis: B (7)

Although downtown Minneapolis does not have a uniformly "historic" district, there are enough interesting, older buildings on the western edge of the core around Hennepin Avenue to constitute a lively nightlife zone. Closer to

Laclede's Landing Historic District in St. Louis

the Mississippi River, old mills and other former industrial buildings have been converted to offices and a hotel. The Upper Lock represents an important bit of functional history next to the falls of the Mississippi River. A good system of plaques encourages visitors to understand the role of the river and its locks in the history of the city. Renovated civic structures plus the old (1928) Foshay Tower add a sense of place to the modern skyline, but the downtown core is largely new and dominated by glass boxes and skyways.

Columbus: B (7)

Downtown Columbus has many nicely restored landmark structures, including the Ohio Theater, the Old Post Office, the state capitol, and the forty-seven-story LeVeque Tower (1927), but overall, so much has been rebuilt that there is little sense of historic continuity. Continuous street walls of older buildings are few and far between. There are, however, historic districts all

around the CBD, making for a "collar" of small lanes and brick buildings that contrast with the spacious core. To the south, the Brewery District and German Village provide a selection of brewpubs and restaurants in a quaint setting, while the Short North/North Market district does the same on the northern edge of the CBD. The emerging Arena District, with its red brick, "industrial chic" architecture, complements the latter. To the east, historic houses give identity to streets such as Bryden Road only blocks from the downtown core, and one area has even been dubbed "Old Town." Across the river, the facade of old Central High School forms the core of the new Center of Science and Industry. The Scioto River, however, is not navigable, and so there are no functional river port activities to add authenticity to the replica of the Santa Maria, which celebrates the name "Columbus" if nothing else.

Indianapolis: B− (6)

Indianapolis has lots of historic civic monuments, including Monument Circle and the state capitol, but little in the way of major historic entertainment districts. The train station has been converted to a festival market and hotel complex, a small version of St. Louis, and there are a few supporting structures nearby. The City Market is also a historic site. Some building facades were saved during the construction of the Circle Center shopping mall and they make for a reasonably interesting streetscape. Most of the historic districts in and around downtown, such as Lockerbie Square, are quiet residential areas, although a section of Massachusetts Avenue (northeast) has been designated a historic commercial strip. There are a number of substantial older buildings in the core that are being utilized for new office and retail uses. New department stores, overflow mall shops like Old Navy, and restaurants now occupy prime street-level space and help to create a nice "urban" feel, at least on a few blocks.

Atlanta: B− (6)

There are a lot of historic sites and districts in and around downtown Atlanta, but for the most part they have not yet played an important part in downtown revitalization. The one exception could be Underground Atlanta, a complex of shops along the old railway tracks below street level, but even it has struggled. It failed during the 1980s and then reopened as a slightly less historic Rouse-designed festival market. A possible future bright spot is the Fairlie-Poplar area, the former downtown core before the city began to march northward on Peachtree. The area is currently experiencing some residential con-

versions. The Martin Luther King/Sweet Auburn historic districts to the east of downtown are important monuments to the struggle for civil rights but are peripheral to the CBD core. The Fox Theater area and other older sections of the Midtown area also add historic character to the city, but they are a long way from city hall. The state capitol and a nice old city hall have less impact than they could, due to their peripheral locations.

Phoenix: C (4)

Phoenix, like Charlotte, is a new city, but it is even more spread out and thus there is even less of a sense of a traditional "main street." The downtown was almost nonexistent until the early 1900s, and so a classic "old town" was never an option. There are some nice Art Deco structures from the 1930s and a few small houses and commercial buildings from the early decades of the twentieth century that have been arranged in a history-themed complex. There are also a few restored theaters and "older" office buildings. Using history as a major downtown attraction, however, has not been feasible. Phoenix has tried to celebrate its heritage by building new versions of "Mexican villages" and the like, but with limited success. There is one residential historic district north of the CBD, and Heritage Square is a minor tourist attraction.

Charlotte: C− (3)

Charlotte is a new city. The metropolitan area contained only 196,000 people as recently as 1950, when it ranked 92nd among urban regions in the United States. As such, it carries very little baggage either good or bad from earlier eras. In addition, in the quest for major city status, most of the older buildings downtown were torn down for new projects. Unlike San Diego, Seattle, or Atlanta, Charlotte rebuilt its downtown on site rather than moving away from it toward a newer zone of assimilation. Much of the area around downtown was destroyed for public housing, a football stadium, and civic space. Nevertheless, there are a few small historic reminders, such as a nineteenth-century shopping arcade and a few old warehouses that are now used for restaurants and galleries. There are also a few Victorian cottages near the downtown core.

Summary: Historic Preservation and Sense of Place

> Seattle 10
> Cleveland 9
> Pittsburgh 9

Baltimore 9
San Diego 9
Portland 9
Providence 9
San Antonio 8
Denver 8
St. Louis 7
Minneapolis 7
Columbus 7
Indianapolis 6
Atlanta 6
Phoenix 4
Charlotte 3

Downtown Housing and Nearby Residential Neighborhoods

Central business districts, according to the classic definitions, are devoid of residential activity. Indeed, with the advent of single-use zoning during the middle decades of the twentieth century, it usually became illegal for people to live in districts that were designated commercial or industrial. Typically, the only people who actually lived downtown were those who dwelled in single-room-occupancy hotels or lived semilegally in rooms over shops or in relic houses on the forgotten margins of the business district. The battle to legalize loft housing in formerly industrial structures in New York City and elsewhere, begun in the 1960s, illustrates this conundrum. While city after city has argued vehemently about the need to get people "back downtown," their own rules and regulations made it very difficult for them to do so. Fortunately, much has changed over the past two decades, and today the provision of downtown housing, both in the form of new construction and the conversion of office and industrial buildings, is a booming business.

Those who wax nostalgic about the old days when large numbers of middle-class people lived downtown are usually referring to the days when people lived close to downtown rather than in it. Indeed, most downtowns were much smaller than the area typically included within the typical "inner belt" definitions of downtown today. Houses and apartments sometimes lined the streets only a few blocks from office towers and department stores, but they were not, strictly speaking, *in* the central business district. Once more, because there are

seldom any consensus definitions of "downtown," it is difficult to say who lived there and who did not during any given time period.

Living Downtown in the 1950s: Life on the Margin

During the decades immediately following World War II, the populations of most downtown areas plummeted. While many authors have referred to this as a voluntary flight to the greener pastures of the suburbs, most of the losses were the result of deliberate city policies. The loss of population occurred in both the core and frame of the downtown. In the core, urban renewal was used to clear old, SRO hotels as well as "nonconforming" mixed-use structures in the name of slum removal. Hotels had been considered an ideal way to live for young, unencumbered downtown workers during the late nineteenth and early twentieth centuries, but by the 1950s, most were considered to be the haunts of unwanted and unemployable "bums." Deliberate policies were aimed at ridding the central city of such embarrassments. Over the next two decades, most American downtowns lost over half of their SRO hotels, often several thousand housing units. Ironically, it was only after these units had been destroyed that civic leaders began to express concerns about the homeless problem. There was no longer any place for very low income people to live.[2]

Housing on the edge of downtown was also being decimated. The residential margins of the central business district offered the cheapest, lowest-resistance routes for freeways, ramps, and interchanges. Inner belt highways tended to increase access to downtown for those traveling long distances by freeway but at the same time cut off downtown from nearby neighborhoods located "just across the road" but with no through streets. For them, downtown was no longer within strolling distance. Meanwhile, the little housing that remained within the beltway ceased to be part of a viable community and was usually zoned commercial and slated for eventual redevelopment. Such neighborhoods were often redlined and experienced disinvestment.

Ghettoization and the Making of the "Inner City"

The urban riots and sense of impending doom that hovered over many American cities during the late 1960s caused residential districts near downtown to be perceived more negatively than ever. In the decades before civil rights legislation arrived during the 1960s, minorities, especially African Americans, were forced to live in well-defined ghettos. Block-busting, steering, and redlining all contributed to the making of high-density, run-down neighbor-

hoods that were typically located near industrial zones and railroad tracks on the edge of downtown. Since there were few if any major retail outlets in these neighborhoods, residents made up a large percentage of the clientele at many downtown stores and coffee shops. As suburban alternatives became more available for whites, the "minority look" of downtown sometimes contributed to white flight from the central city. The fact that many downtown stores and entertainment districts appeared worn out after decades of disinvestment also played a part in the perception of urban decline.

Urban renewal and the construction of high-rise, low-income "projects" in and around downtown accentuated the negative image of central city living. They also changed the character of some of the remaining pleasant neighborhoods close to the CBD. Stunning images of bleak towers, old houses, boarded up commercial strips, and abandoned cars appeared regularly in the American media during the 1960s and 1970s. The demand for middle-class downtown housing was, to say the least, not very great.

Apartment Districts and Urbane Living: The Struggle to Hold On

Prior to the 1960s, very few American cities had the kind of large, luxurious apartment buildings that made central city living so popular in cities such as

Public housing on the edge of downtown Baltimore

Paris or Buenos Aires. New York City, Washington, D.C., San Francisco, and Chicago all had extensive apartment zones within walking distance of the downtown core by the 1920s, but beyond this short list, such neighborhoods were few and far between. As late as 1960, Manhattan contained as many housing units in large apartment buildings (ten or more units) than the next twenty-three largest American cities combined.[3] Most Americans wanted to live in free-standing houses, and that usually meant moving further away from downtown. This was especially true in industrial cities such as Detroit or Cleveland, where even apartment buildings, especially the new garden apartments, were built well away from the smoky urban core.

When apartment construction resumed in the late 1950s, after the hiatus of the Depression and World War II, most units were in low-rise suburban complexes. Some were built in small, in-fill projects around universities and employment nodes, while others were built around amenities such as golf courses, lakes, or tennis clubs. Downtowns saw very little construction. Meanwhile, many of the older mansions located around downtown, often in mixed commercial and apartment zones, fell into disrepair as the costs of deferred maintenance mounted. The prospect of adding modern plumbing, wiring, insulation, kitchens, etc. to massive Victorian-era houses was scary enough even in "good" areas, but in the neighborhoods around downtown, it just did not seem worth the trouble and expense, especially if loans were hard to get and taxes were high.

Housing Comes Back: Gentrification, Conversion, and New Construction

By 1970, very few people lived in most American downtowns, and only a minority of major cities had stable, economically viable residential areas adjacent to the CBD core. But things began to change in the 1970s. The "urban crises" of the 1960s demonstrated that something needed to be done. Also, as suburban housing prices began to escalate rapidly during the mid-1970s, it became clear that much of the "inner city" was undervalued and that there were deals to be had and profits to be made. The "back to the city" movement that has impacted many American urban centers over the past twenty-five years consists of three separate but related components—the gentrification of older, residential neighborhoods, the conversion of industrial and commercial structures into residential "lofts," and the construction of new condominium and apartment complexes. The first of these to have an impact was gentrification.

By the early 1970s, a new appreciation of older housing with "historic" character arose and was encouraged by changes in tax laws and by projects associated with the fast-approaching Bicentennial celebrations in 1976. Gentrification refers to the filtering up of existing residential areas so that each new wave of residents is higher status (in terms of income, education, etc.) than the last. This process runs counter to the classic filter-down models of neighborhood change associated with, for example, the concentric zone model of city structure. Many interesting metaphors were used to describe the early stages of the process, as "urban pioneers" moved into the "urban jungle" and established "beach heads" for further "revitalization." In such targeted neighborhoods, low-income residents were often displaced involuntarily as rents and taxes rose. Many interesting microgeographies unfolded, as when, for example, middle-class pioneers located so as to use tough, white, working-class areas as buffers against ghetto expansion. Conflicts were inevitable.[4]

Not every city experienced gentrification. The best candidates for the process were cities with strong downtown employment bases and at least a few remaining desirable neighborhoods close to the core. Gentrification thus had a major impact in cities such as San Francisco and Boston but was much less important in Detroit and Newark. Most cities were somewhere between these extremes. Cities with strong white ethnic neighborhoods often attracted lots of gentrifiers, since a colorful ethnic identity could be used as a theme in the creation of alternatives to suburbia. "Greek towns" and "German villages" were "sold" as being interesting and lively places compared to bland suburban tracts. Cities dominated by Latino barrios and African American ghettos had a tougher time attracting "back to the city" residents, but even in these cases, there has been some turnaround. Cities with large industrial zones encircling the downtown are least likely to experience residential gentrification since there are few neighborhoods to upgrade.

There are many pros and cons to gentrification. How much is enough? How much is too much? Housing prices in central Boston and San Francisco are now astronomical, and much of the life and color of the cities is disappearing as more and more neighborhoods are dominated by professional people who work long hours in offices. Dot.com Village may not be as colorful as Little Italy. On the other hand, central cities that have experienced little or no gentrification lack the threshold populations for almost any kind of economic revival and are losing both people and businesses as the building stock continues to deteriorate.

By the 1980s, the conversion of old warehouses, factories, wharves, and breweries to "loft housing" had become the rage in many American downtowns. For decades, such structures had been strictly off limits for residences since they were zoned commercial or industrial. Strict adherence to building codes also meant that even if the zoning was modified, the structures would have to meet all the current code requirements for residential buildings. This was usually difficult or even impossible to accomplish economically in old warehouses. Over the past two decades, a variety of innovations in flexible multiuse zoning and performance standards for fire and seismic codes have helped to facilitate the conversion of tired commercial buildings into residential units. In some cities, even older, first-class office buildings have been converted to apartments and condominiums.

The conversion of industrial buildings often involves a substantial cleanup of toxic materials. Most American downtowns have had to deal with environmentally problematic "brown fields" as they have expanded outward into formerly industrial zones. Once again, flexible interpretations and solutions have been worked out in many cases, but controversies remain. Still, artists' lofts have become so popular in many downtowns that a substantial number of new "loft buildings" have gone up. For those seeking an alternative to suburban housing tracts, "industrial chic" may be just the thing.

The biggest potential source of housing units located in the CBD core, as opposed to the frame, is new construction. As long as Americans insisted on living in single-family houses, the prospect of building new housing downtown was nearly nonexistent. Only in places like Midtown Manhattan, where people were willing to live in massive residential buildings, could housing hold its own against "higher uses" such as office buildings and department stores. In recent years, a significant number of Americans have decided that apartment living is a viable option and large buildings have been built in many locations, including downtown and in suburban edge cities.

Since living downtown was a novel idea for most middle-class Americans, at least during the initial stages of construction, and because many downtowns still had an empty, even scary, feeling at night, most early residential buildings were designed like fortresses. Larger structures had in-house, usually underground, parking and locked and gated entryways. Very rarely did doors open directly onto the street, and blank walls predominated. Mixed use was also rare through the 1960s and 1970s, and so retail and service activities were not normally used to enliven the lower floors. These self-contained structures did not

Loft housing in the LoDo historic district in Denver

contribute much to a sense of downtown urbanity since there were rarely any people outside. As housing has become more common, acceptable, and even normal, more new residential buildings have taken on a softer, more street-oriented appearance. This is especially true in districts that have become heavily residential, as opposed to business districts with an occasional apartment tower.

Today there are two main types of new downtown housing: high-rise buildings in or very close to the core and low-rise and townhouse developments on the fringe. The former type is still relatively rare in middle-sized American downtowns. Of the cities in this study, only San Diego and Seattle have many structures in this category. Residential buildings still have trouble competing (bidding) for prime downtown locations, and many potential residents find the idea of living in a financial district unappealing. On the other hand, smaller cities do have an advantage over major urban areas such as New York or Chicago because their financial districts are much smaller and less specialized. In addition, the trend toward mixed-use, high-rise buildings that include office, residential, retail, and hotel space are beginning to appear. If successful, such structures may play a major role in reshaping the future of American downtowns.

High-rise luxury housing in the core of downtown Seattle

So far, new low- and mid-rise structures on the edge of downtown have played a far more important role in the creation of downtown residential populations than have towers in the core. The two most common types of edge districts are those that focus on some major amenity such as a waterfront or park and in-fill zones that are located between the downtown core and existing residential areas so as to form a continuous residential sector. Revitalized waterfronts have acted as major foci for housing construction in many American cities. They offer a "place" that not only is worth being in but also increases the possibility of having jogging trails, bike paths, restaurants, and other amenities nearby.

While the two are not mutually exclusive, sector in-fill housing illustrates a different dynamic. Here, the attraction is related both to access to the downtown core and identity with a long-established desirable neighborhood. Thus,

for example, new residential towers near Chicago's Navy Pier offer both views of a revitalized waterfront and a connection to the traditional Gold Coast residential sector. Obviously, cities that no longer have reasonably attractive close-in neighborhoods have a tough job creating residential sectors. In such cities, new downtown housing is destined to remain an isolated island, connected to the downtown core but to little else.

Gentrification and "back to the city" or at least "back downtown" movers have received most of the attention recently, since they represent people who live downtown by choice. Many of those who lived downtown before about 1980 were often there only because of poverty or discrimination. They had no choice and often left when options became available. But new residents are not the only players in downtown residential viability. In many near downtown neighborhoods, incumbent upgrading has become very important in recent years. This implies that long-time residents with options voluntarily stay and participate in maintaining and upgrading their homes. This may occur in white, white ethnic, Latino, Asian, or African American or culturally diverse neighborhoods with a wide variety of income groups and housing prices. It has been very difficult to achieve this kind of mixing in most American central cities. If prices and levels of maintenance get too low, no one will invest and the area gradually becomes uninhabitable; if prices get too high, excessive gentrification occurs as speculative development takes over. The trick is to spread out demand so that new investment is not overly concentrated in a few trendy zones but rather acts as a catalyst for the gradual upgrading of several neighborhoods simultaneously.

In many American cities, space for downtown residential revitalization has become available not only because of deindustrialization but also because of the declining number of new, low-income immigrants seeking housing in and around downtown. No longer are people from Europe, Appalachia, or the Deep South pouring into the central cities of the American heartland. Rather, many of them have abandoned their initial neighborhoods and moved to suburban locations. But this is not true everywhere. In much of the Southwest, for example, huge numbers of immigrants from Latin America, Southeast Asia, and Africa have recently sought housing in central cities. Certainly a major variable differentiating American cities today is whether or not traditional immigrant districts are still forming in and around downtown.

While the total number of residents in most American downtowns is still quite low compared to sprawling suburban areas and most people are still seek-

New housing along the canal walk in Indianapolis

ing single-family houses with green lawns, downtown options are now appearing in greater numbers than ever before. A few middle-sized downtowns now have as many as twenty thousand inhabitants, and the number may double in the next decade. This is significant for a number of reasons. Not only does it demonstrate that downtowns can become twenty-four-hour places with a variety of activities, but it also shows that Americans who have a choice may be enticed into living at higher densities. As the problems associated with sprawl and auto-dependency become increasingly obvious, it has become clear that something has got to change.

Below are my evaluations of the successes and failures of American downtown residential areas. The population figures I include are from the 2000 census, but of course downtown estimates can vary widely depending on the tracts included in a downtown definition. Population figures, therefore, must always be considered approximate. In addition, I also consider quality and variety in my rankings of good residential downtowns. Places with a few large but isolated towers are not, in my opinion, as "good" as places where housing is well integrated into the life of downtown.

Variable 9. Downtown and Near-Downtown Residential Areas

San Diego: A (10)

San Diego has one of the most heavily populated middle-sized downtowns in the country. In its attempt to revitalize what was a pretty moribund downtown, San Diego has emphasized the construction and rehabilitation of a wide variety of housing and neighborhood types. The Marina District, a former warehouse area near the waterfront, has been zoned residential and now contains luxurious condominium towers, new "loft" structures, single-room-occupancy hotels and subsidized housing for the elderly. There are also a number of low-rise, market-rate apartments and condominiums of all sizes and shapes. In Cortez Hill on the opposite side of downtown, a 1928 hotel has been converted to luxury apartments and is being surrounded by a wide range of housing options. There are also new apartments and converted lofts in the Gaslamp Quarter and in Little Italy. There is now an estimated population of about twenty thousand within the downtown, which is enough to support the new Ralph's supermarket. Many low-income residential hotels have been destroyed, however, and homelessness is a serious problem, especially on the eastern fringes of the CBD. Plans call for the construction of more low- and moderate-priced housing there, but a concentration of homeless shelters and other poverty-linked social services has deterred investment. Now that downtown has been established as a desirable residential location, however, the pressure is on for something to happen.

There are also excellent, walking-distance connections to vibrant neighborhoods to the north and east of downtown. To the east, Golden Hill is a diverse and growing neighborhood with interesting mixtures of housing and people. It remains one of the city's primary barrios. To the north, Balboa Park affords an attractive amenity for new condominiums and apartments, even if much of it lies under the flight path to the city's main airport.

Seattle: A (10)

Downtown Seattle has the disadvantage of having a massive industrial zone to the south as well as the waterfront to the west, and so linkages to nearby residential neighborhoods are limited. Only to the east toward First Hill is there a neighborhood with a variety of housing options from mid-

High-rise condominium towers in downtown
San Diego

rise apartments to older, restored houses. Downtown construction, however, is booming, especially in the Belltown district just to the north of the
CBD core and along the northern waterfront. Pioneer Square and Pike Place
Market have become major foci for residential construction and rehabilitation as well. Some low-cost SRO hotels remain, but their numbers have declined. Although many residences were destroyed to create the Seattle Center in the 1960s, there are a number of viable neighborhoods just beyond it.
Views of the skyline and the bay enhance the appeal of many close-in housing options. The downtown core now has a residential population of about
twenty-four thousand, and it is the fastest growing neighborhood within the
city limits.

Portland: A (10)

Downtown Portland is gradually being surrounded by housing, at least on three sides. To the south, apartments created as a result of urban renewal in the 1960s are being joined by new projects along the Riverplace Marina. To the north, the Pearl (industrial) District is experiencing both new construction and loft conversion. To the west, a very mixed but rapidly gentrifying residential area focusing on Twenty-third Avenue is the only near-downtown area with a traditional residential neighborhood atmosphere. It also helps to have some very wealthy residential areas in the hills overlooking downtown, since the "movers and shakers" there have always tended to see downtown as their shopping and entertainment district; this probably helped to keep downtown alive during the struggling postwar decades. Low-cost shelters and SRO hotels remain in the Old Town area, but it will probably be hard to maintain them in the face of continuing housing price inflation. The downtown core has a population of about thirteen thousand, and there are many more in the nearby hills.

Minneapolis: A− (9)

Downtown Minneapolis has attracted large numbers of residents (about twenty thousand) to apartment districts both to the north and south of the CBD core. To the south, Loring Park and the Walker Art Center provide attractions for new mid- and high-rise apartment construction. The district blends into an area of older houses and apartments further south and west. There are also new residential towers along the Mississippi River to the north of downtown. In addition, a variety of older houses and apartment buildings have been renovated on Nicollet Island. The huge campus of the University of Minnesota lies to the east of downtown and some large residential projects have gone up around it. Connections with the downtown core, however, are not great. Good access to close-in traditional residential neighborhoods exists only to the south.

Baltimore: B+ (8)

Downtown Baltimore has relatively little housing in the core, but there are at least four interesting neighborhoods nearby. To the east, Little Italy contains a wide variety of old row houses, public housing projects, and new residential

Housing towers near Loring Park in Minneapolis

structures. To the south, the historic Federal Hill area has experienced considerable renovation as well as a new residential tower, but it is still a very mixed neighborhood. To the northwest, African American neighborhoods abut the Lexington Market area and use downtown as a major shopping destination. In addition, some restored houses and apartment buildings exist in the Mount Vernon District around the Washington Monument on North Charles Street. To the southwest and northeast, however, major highways cut off the downtown from the neighborhoods beyond. Much of the nearby "inner city" is either in marginal shape or contains large public housing projects. This means, however, that gentrification and high housing costs will not have the same problematic impact as they do in, say, Portland or San Diego. Baltimore has about sixteen thousand people living downtown, with nearly as many living in neighborhoods not far away.

Denver: B+ (8)

Downtown Denver has a large and increasing stock of loft apartments in both new and old structures in the LoDo district, but there are few other residential "neighborhoods" downtown. East of the state capitol, the Swallow Hill

Renovated row houses on the edge of downtown Baltimore

district contains a variety of older houses and apartment buildings in a sector heading out toward City Park, and new multiunit structures have been constructed along Cherry Creek to the southeast of the CBD core. The vast empty fields that separate downtown from the South Platte River and the neighborhoods beyond are slated for housing developments in the not too distant future. There are some lower-income residential areas along the industrial corridor to the northeast but they are not well connected to the core. Some low-income residential structures were destroyed in the big urban renewal projects leading to the convention center and Auraria campus, and that further separates downtown from districts with a residential ambience. The greater downtown area has about fifteen thousand residents.

Columbus: B (7)

Columbus has relatively little housing in the CBD core, although there are still a few streets with historic houses and low-rise apartment complexes. However, downtown is nearly surrounded by a variety of close-in residential neighborhoods. There are two high-rise condominiums and several apartment buildings on the southern edge of downtown close to the historic German Vil-

The German Village historic district next to downtown Columbus

lage/Brewery District. Further south, German Village itself is a large neighborhood of restored houses. To the north, housing in the emerging Arena District will soon enhance the largely residential Victorian Village and Italian Village neighborhoods. To the east, in a largely African American neighborhood, some "Rainbow Coalition"–style renovation has taken place, leading to greater diversity but not excessive gentrification. Social services and homeless shelters are concentrated on the west side of the Scioto River, and some low-cost housing remains there as well. Although they are beginning to appear, there are few loft districts or converted residential structures downtown. It is possible, however, to stroll into a variety of traditional residential neighborhoods only minutes from the CBD core. About seven thousand people live in the greater downtown area, with another eight thousand close by.

Providence: B (7)

Almost no one lives in downtown Providence, but the tiny CBD is surrounded by residential communities on the hills just beyond. Many of these are within easy walking distance of the core and are visually important to the overall sense of place. College Hill, just to the east, is filled with a variety of mansions, apartments, and student housing, while the west side features a working-

class ambience with some gentrification. There is also housing just beyond Capitol Hill. These communities are much closer to the core than they would be in typical, more spread-out downtowns. Only about forty-three hundred people live downtown, but there are nearly twenty thousand more in nearby downtown-oriented neighborhoods.

Indianapolis: B− (6)

Historically, downtown Indianapolis has had neither housing nor good connections to nearby residential neighborhoods. The area south of the downtown railroad tracks is largely industrial and walls it off from the neighborhoods beyond. To the northwest, much (largely African American) housing was destroyed to create the Indiana-Purdue University campus. Only Lockerbie Square, a very small and partially rebuilt residential area to the northeast, provides a downtown historic housing option. There are also a few housing towers dating from the 1960s. However, things are improving. New apartments and condominiums have been built along the canal walk to the west of the CBD core and could bolster gradual revitalization and in-fill in the nearby African American community. Ironically, success in other areas of redevelopment (such as sports complexes and universities) and the continuing viability of the Eli Lilly industrial complex may have limited the number of close-in areas that are suitable for residential development. Since the inner belt is far away from the core, the city features a rather large and amorphous "greater downtown." It houses about twelve thousand people.

Atlanta: B− (6)

To a very real degree, downtown Atlanta is divided into a black south and a white north. As a result, much of the southern zone has experienced a combination of urban renewal for highways, stadiums, and government buildings while also experiencing a minimum of new privately financed housing developments. To make matters worse, over the years these areas have been filled with very low income "projects," many of which are badly deteriorated. On the other hand, the northern "Midtown" zone around Piedmont Park has become a trendy area for upscale apartments and condominiums. Between the two extremes, however, some interesting things are taking place. Inman Park, an old neighborhood close to the racial "border" has been discovered as a good area for restoration, and several others, such as Cabbagetown, are not far behind. Neither, however, is within walking distance of the CBD.

The old downtown core, Fairlie-Poplar, is experiencing some conversion from office to loft residential. The new Centennial Park created for the Olympic games may also serve as a focus for some new residential activity. The nearby Techwood projects have been rebuilt as a mixed subsidized and market-rate residential area of townhouses. There are only fair connections between most neighborhoods and the downtown core, however, and so the whole remains somewhat less than the sum of its parts. It is possible that the Sweet Auburn area east of downtown may have some appeal for African American urbanites, but little has happened as yet. Downtown Atlanta has about eleven thousand residents.

Pittsburgh: C+ (5)

There is very little housing in Pittsburgh's small and compact downtown, although some loft housing is beginning to appear in the Cultural District to go along with the small projects that resulted from the redevelopment of the Point during the 1950s. Perhaps the best example of close-in residences is the new, neotraditional complex of Crawford Square just to the east of the CBD in the largely African American "Hill." Beyond these examples, the big problem in Pittsburgh is the topographical difficulties involved in linking downtown to nearby neighborhoods. Dusquesne Heights and Washington Heights loom above the urban core and offer wonderful views of the skyline, but connections are awkward even with the fun-to-ride funiculars. The combination of rivers, hills, canyons, and remnant industrial zones makes it difficult to build much housing within walking distance of downtown. On the other hand, many neighborhoods still rely on downtown for shopping and entertainment, due to the continuing importance of public transit and the difficulty involved in getting to suburban options. About twelve thousand people live in and around downtown Pittsburgh.

San Antonio: C+ (5)

A few older office and warehouse structures have been converted into housing and there are a few small apartment complexes along the Riverwalk, but in general there is little housing in downtown San Antonio. The fine old houses in neighborhoods such as the King William Historic District are picturesque but contain few units. A bleak housing project just to the south of Hemisfair Plaza is being redeveloped and should link up with an emerging gallery district as a residential focus. A modest but picturesque neighborhood extends toward

the northeast and the renovated art museum, but there is little sense of its being downtown. Freeways and the massive Alamodome cut off some interesting neighborhoods to the east of the CBD, while highways and industry do the same on the west side. About six thousand people live downtown.

Charlotte: C+ (5)

There is a nice neotraditional neighborhood not far from Charlotte's CBD core, along with some new apartments and condominiums, especially in the new Gateway Village complex. A bleak, semiabandoned public housing project surrounded by empty fields has recently been rebuilt with moderate-density housing on Ninth Street. There are also some very nice neighborhoods beyond the inner belt but not really within walking distance of downtown. There is plenty of space and lots of potential for downtown living, but things have only just begun. The combination of many pleasant "suburban" residential options within a few minutes' drive of downtown and the lack of any real historic neighborhoods in the center city makes for a relatively weak demand for downtown housing. About six thousand people live in the greater downtown area.

Neotraditional housing in downtown Charlotte

St. Louis: C (4)

St. Louis has been trying to bring housing downtown for several decades, but the cards are stacked against its success. Over the years, the central city has been ravaged by ghettoization, disinvestment, and deindustrialization and remains full of vast empty fields, bleak public housing towers, and abandoned buildings. Creating a sense of amenity has been difficult. Nevertheless, there is a huge stock of fine old buildings that are gradually being utilized for loft apartments, and the potential is there for a lively downtown neighborhood. Some fairly new (1960s) apartment buildings remain from the days of urban renewal, but there have been few new projects in the CBD core. Some of the worst public housing towers have been torn down and replaced with subsidized but pleasant townhouse developments, but dereliction is still widespread. Although housing renovation has occurred in the Lafayette Square and Soulard neighborhoods, the areas are small and poorly connected to the CBD core. Approximately sixty-five hundred people live downtown, but there is still much nearby abandonment.

Phoenix: C (4)

Downtown Phoenix is surrounded by very low density areas, with vast open fields awaiting (perhaps) eventual development. There are some high-rise apartment buildings on Central Avenue to the north, but they are located well away from the CBD core. Some newer, low-rise apartment complexes have been constructed on the margins of downtown recently, but there is as yet little sense of a downtown community. The major achievement is the addition of new mid-rise housing in the Roosevelt Historic District north of downtown. The new park covering Interstate 10 acts as a pleasant attraction for housing. The potential for more residential in-fill is there, but it will probably be a while. The census tracts are very large and include many areas with single-family houses, making it difficult to identify "downtown" residential areas. About six thousand people live downtown.

Cleveland: C− (3)

Downtown Cleveland is experiencing a boom in loft housing and new apartments in the Flats and the Warehouse District, but that is about it. Vast industrial (and former industrial) zones separate the CBD from the residential city on two sides, and Lake Erie forms a neat edge on the third. A ghettoscape

Apartment complex on the edge of downtown Phoenix

of commercial, industrial, and residential abandonment predominates for several miles to the east. Downtown is all by itself, with few good connections to
the neighborhoods beyond. Lofts are the only game in town. About five thousand people live downtown.

Summary: Downtown Residential Neighborhoods

San Diego 10
Seattle 10
Portland 10
Minneapolis 9
Baltimore 8
Denver 8
Columbus 7
Providence 7
Indianapolis 6
Atlanta 6
Pittsburgh 5
San Antonio 5

Charlotte 5
St. Louis 4
Phoenix 4
Cleveland 3

Transit Options Downtown: Light Rail, Bus Corridors, Bike Paths, and "Autoscapes"

How many cars can a downtown hold and still have an urbane, pedestrian-oriented atmosphere? Over the past half-century or so, most American downtowns have been redesigned in order to accommodate the automobile. Streets have been widened, buildings have been torn down and replaced by parking lots and garages, and massive freeways and interchanges have encircled downtowns and separated them physically and psychologically from the residential areas beyond. Unlike the situation in Europe, where outer belts are common but inner belts in the core are rare, Americans have built freeways throughout central cities. Some freeways have been built below grade so as to have less impact, but others have been constructed at or above street level and dominate the landscape. The roar of the traffic can be deafening in nearby buildings. In addition, the elimination or reduction of on-street parking in many downtowns, along with an emphasis on one-way thoroughfares, has turned existing streets into semifreeways with fast-moving traffic rather than "stop and shop" auto usage. Massive ramps and parking garages, along with vast fields of parked cars, have diluted most American downtown environments. Walking is a bleak experience in such landscapes.

As if to add insult to injury, downtown buildings often have been designed to cater to automobiles rather than people, especially at sidewalk level. Underground and street-level parking in office, hotel, and residential towers typically involves gaping entryways with wide curb cuts through the sidewalk. Strolling along a street is no fun when cars are barreling in and out of cavelike garages. The blank walls that dominate most parking structures also dilute the urban experience. On the other hand, not everything built for cars and trucks is necessarily ugly. Some very beautiful bridges span the waterways of American downtowns and some lovely "great streets" and boulevards exist. Still, it is easy to argue that automobiles have had a largely negative impact on most downtowns. In this section, I attempt to evaluate this negative impact as well as the increasing number of transportation alternatives.

The Alaskan Way Viaduct between downtown Seattle and the waterfront

Downtown Spreads Out

Compared to the downtowns of the pre–World War II era, today the typical city core occupies a much larger territory. The monumental scale of convention centers, sports arenas, shopping malls, and housing complexes, along with streetless superblocks from the days of modernist planning in the 1950s and 1960s, means that getting from one spot to another can be difficult. If the distance from one attraction to another is over a mile and the linkages are awkward or unpleasant, the downtown may not hold together as a single place. Rivers, hills, and other topographical features that may add visual interest to a downtown can also act as barriers to the easy flow of people and activities.

The tendency of American downtowns to expand outward along a major spine toward a zone of assimilation can both spread the CBD too thin and make walking impractical. On the other hand, a linear downtown can facilitate the use of light or heavy rail transportation with nodes of activity near the stops. The geography or spatial organization of a particular downtown can be just as important as the type and number of activities located within it. If each node is isolated and disconnected from the others, downtown can become little different from a suburban edge city strung out along a major highway.

The Uneven Revival of Mass Transit

The construction and expansion of urban rail systems ground to a halt during the Depression in all but a few large cities like New York. With the automobile-dependent suburban boom that began during the late 1940s, transit alternatives were all but forgotten. Streetcar and light rail systems, the most famous being the Big Red Cars of Los Angeles, were abandoned and highways were widened. After the passage of the National Highway Act in 1956, the shape of the typical American metropolitan area was increasingly determined by the location of omnipresent freeways. By the late 1960s, however, it was clear that all of the eggs could not fit in the freeway basket. A few large urban areas such as San Francisco and Washington, D.C., began to ponder the wisdom of building subways. By the 1970s and 1980s, many new heavy (subway) and light rail systems were under way in cities all across America from Baltimore to San Diego.

In most middle-sized urban areas, mass transit is controversial. Some have argued that it is absolutely essential that we begin to work toward creating sustainable cities with transit-oriented development.[5] They suggest that the automobile era was a brief one and that sprawl and congestion are now approaching the limits of acceptability. In only fifty years, automobiles have gone from being expensive toys to ubiquitous modes of transportation. Everything from taking the kids to school to buying a loaf of bread requires a trip in the car. The system will someday collapse under its own weight. On the other hand, many argue that mass transit, especially rail transit, is an expensive boondoggle and that people will continue to drive and to demand better highways and more parking; for those who cannot drive, a good bus system is all the mass transit that American cities really need.

Opponents argue that rail transit is the transportation equivalent of skyscrapers and new baseball parks, meaning they are fancy status symbols that give "big league" identity to a city rather than an essential service. Mass transit may even encourage sprawl by making long-distance commuting easier. Federal subsidies for transportation innovations are seen, like urban renewal funds years earlier, as free goodies to grab while they are available, whether they're needed or not. The enthusiasm for innovative transportation has been uneven. Some cities have gone all out to focus development around mass transit while others have continued to emphasize the automobile. Some cities have tried to develop both.

Balanced Transportation and Urban Aesthetics

Ever since cities became large, aesthetically marginal transportation facilities have taken up huge amounts of space. Smoky ports and rail yards dominated most American downtowns until the mid–twentieth century. A "Chinese wall" of tracks cut through the center of Philadelphia and rail lines took over much of the waterfront in Toronto and Chicago. Steamboats, ships, and warehouses dominated the riverfronts of Cincinnati, New Orleans, and St. Louis. Getting people and stuff in and out of the central city has always been a major urban design problem. Elevated commuter lines cluttered the streets of New York, Boston, and Chicago, while trains at street level sliced through nearly every American city. Subways, while less visually intrusive, were extremely expensive to build and caused a great deal of chaos and disruption during construction. During recent decades, massive freeway interchanges and elevated highways have become the major disamenities in typical American downtowns, taking over waterfronts and slicing through hillsides.

As downtowns have matured, serious efforts have been made to diminish the negative aesthetic impact of transportation. Rail yards have been removed (Chicago) or covered over (Philadelphia) and waterfronts have been cleaned up for recreational uses. The most interesting efforts, however, have involved highways. Portland, Oregon, removed a highway from its waterfront and Boston is in the process of undergrounding a downtown freeway. San Francisco, with the help of an earthquake, got rid of the elevated highway along the Embarcadero. Seattle built a park over a downtown freeway and Phoenix followed suit. Open "fields" of parking are gradually being built upon, and parking garage design is becoming at least a little more architecturally sensitive. In addition, a few cities are reviving the idea of "head in" parking, both to increase access to local stores and to slow traffic on downtown streets.

Given all the controversies, the ideal would seem to be a pleasant balance of transportation options rather than simply pushing one form, such as rail, over another, such as freeways. A good downtown might have grand boulevards full of traffic, bike paths, pedestrian promenades, light rail systems, bus malls, commuter rail stations, and limited access highways. The problem is not the automobile. There are plenty of cars and traffic jams in European cities, but urban planning and design there does not simply revolve around making space for the car. In American downtowns, however, that has too often been the case. For years, downtowns have been decimated as buildings have been cleared and

streets widened in an effort to get more cars into the city. Since most cars are driven only a few hours per week, storage is a big problem. Parking lots often take up more space than any other land use.

Defining the Perfect System of Downtown Transportation

The transportation facilities designed to bring people into downtown should be located on the edge of the CBD core rather than within it. This seems like a simple idea but it is often difficult to implement. Many downtown parking lots were created during the 1950s and 1960s through the random clearance of "obsolete" buildings. Their location, therefore, is somewhat scattered and ad hoc. On the other hand, many garages have been built adjacent to or under major shopping and employment centers and so are as close to the middle of the CBD as possible. Although some centrally located garages are essential, as these structures have gotten larger—many holding between three and five thousand cars—they take up a lot of space and contribute mightily to congestion, especially at closing time. A possible solution is to locate massive garages on the edge of the CBD and connect them to shopping centers and office buildings via both street-level pedestrian zones or second-level walkways. These paths can be lined with coffee shops, card or flower stores, restaurants, and a variety of colorful amusements aimed at generating economic activity beyond the few major buildings of the downtown core. As drivers exit these garages, there should be plenty of options, including freeway ramps and city streets heading in a variety of directions. Too often, parking garage exits channel thousands of cars onto a one-way street leading to a freeway, causing not only congestion but also a solid barrier of cars to those hoping to go elsewhere. Major bus and rail terminals should be located similarly.

Once they are downtown, people should have the opportunity to use a variety of types of transportation to move around within it. Of these, the one that is being pushed most actively is walking. Joel Garreau has maintained that people do not want to walk more than about six hundred feet.[6] If his estimate is correct, designing a walkable downtown is an uphill battle. I would argue, however, that how far people are willing to walk is a function of the pleasures and obstacles they encounter along the way. Most people can walk at a twenty-minutes-per-mile pace. If this is so, going a half-mile should be no problem for a lunch break or an after work stroll. For those who like a brisk pace, a fifteen-minute, one-mile walk is also a reasonable option. Of course, if the landscape is bleak and there are awkward linkages, people are less likely to wander.

Two innovations aimed at encouraging walking in American downtowns have had limited success. The first is the pedestrian mall. While pedestrian streets have worked well in Europe where the population densities are higher and the reliance on small-scale, street-level shops is greater, they have not had the same success in American cities. Many quickly became dead and semi-abandoned when cars were removed, especially if they were longer than one or two blocks. Slow-moving traffic and on-street parking seem to add a necessary colorful element to lively American streets.

The second problematic element is a system of second-level walkways providing indoor connections between buildings. Pioneered in Minneapolis and Calgary, where weather can be a major disamenity, they have diffused to a variety of other cities. Whereas they can be useful for connecting, for example, peripheral parking garages to CBD office buildings and shopping centers, they can also be overdone. At worst, they disrupt street vistas and contribute to empty sidewalks and underutilized front doors. Perhaps the best way to encourage people to walk around downtown is to provide lots of street-level attractions, window-shopping opportunities, inviting doorways, pleasant vegetation, street embellishments, and a sense of security.

Downtowns have also experimented with a number of other ways to get people to explore their environs. Free downtown bus zones, such as Seattle's Magic Carpet Ride and Portland's Fareless Square, encourage people to hop on and hop off at various sites without having to worry about having the proper change. A bus mall, or semipedestrian street with buses only, can also be quite effective. Since many American downtowns have a strong linear dimension, a parade of buses going from one end to the other usually works well. Cities have also tried a variety of free or "by the day" shuttles that stop at major attractions and shopping venues. Some of these are designed to look like colorful cable cars or train engines in order to make moving through the city seem pleasant and entertaining. Light rail systems have also been built to accentuate the peak land value intersection and other nodes. Their permanence and predictability encourages transit-oriented development.

Cities with a strong sense of center can enhance that centrality by having a number of bus lines come together in one place. On the other hand, if a central plaza becomes little more than a giant bus terminal, the downtown ambience can easily change for the worse. Still, a central point where everyone knows that buses to and from everywhere converge and where good, clearly presented information and advice is readily available can make moving around the city

A pleasant streetscape at a downtown Portland bus stop

easier. Attractive and sheltered bus stops on streets with lots of spur-of-the-moment shopping opportunities, especially coffee and snacks, can make waiting for a bus or trolley a lot more fun. Bleak and dreary waiting areas are sure to kill any transit system.

A few cities have experimented with more exotic alternatives, though few of these have had much impact. Monorails and elevated "people movers" have appeared from Miami to Seattle, but they are expensive to build and maintain and are often aesthetically problematic. Of greater importance for everyday use may be the old standby, boats. In San Francisco, Seattle, and San Diego, ferries still bring people into the downtown from residential suburbs and nearby hotel districts. In Boston, visitors can take a boat into downtown from the airport, and the Staten Island Ferry to Lower Manhattan still provides a pleasant (and free) commute. In San Antonio, boat rides along the Riverwalk are a major tourist attraction. Now that there are tour boats on the once flammable Cuyahoga River in downtown Cleveland, anything is possible.

There are also historic transportation systems. Once widely used throughout America, cable cars have held on only in San Francisco. On the other hand, horse-drawn carriages and even bicycle versions of the rickshaw are becoming

A funicular to Washington Heights above downtown Pittsburgh

more common, especially where tourism thrives. As riverfront parks are expanded and bike paths are introduced on secondary streets, cycling has become an option for the energetic. Couriers have long used bicycles in even the biggest cities to deliver packages in a hurry, but the effort is not for the faint of heart. Still, the more transit options there are, the more likely people will be to explore downtown.

Spatial Organization and the Perception of Distance

The success of transportation downtown depends to some extent on not only the objective distances between destinations but also the perception of distance. People will walk, take a cab, bus, or trolley to places they think are reasonably close and easy to get to, but they are less likely to travel to mysterious and disconnected locations. Different downtowns may occupy similar amounts of total territory but have very distinctive spatial patterns. In some cases, related attractions such as major hotels, convention centers, sports facilities, and museums may be very well linked, while in others, they may occupy separate corners of the CBD. In some cities, downtown activities are distributed along a walkable linear corridor enhanced by a light rail system or bus

mall, while in others, dreary or even scary empty zones lie between potential attractions. Downtowns that are imageable—that is, easy to picture and mentally map—are more likely to be explored than those that are confusing. Good transportation systems, therefore, must include clear signage and information. Good downtowns should have both major landmarks and clearly understood pathways and corridors.

It is difficult to give high scores to American downtowns when it is obvious that they are far behind the cities in Europe and Asia in developing comprehensive and efficient multimodal transportation options. American cities still depend on the automobile and auto-dominated landscapes that have scarred and disrupted most of our downtown areas. Nevertheless, there has been progress on many fronts and the scores below represent what is possible in present-day America rather than the state-of-the-art systems associated with, say, Japan.

Variable 10. Downtown Transportation Options and Associated Landscapes

Portland: A (10)

Portland, the first city to remove a freeway from its downtown, has developed an excellent and varied transportation system. The downtown core includes a bus mall with pleasant, covered bus stops all along the way. The mall interacts with a light rail system in the heart of the retail and entertainment district. The downtown "Fareless Square" district encourages people to ride transit for short distances. Street-level retailing and a canopy of trees make waiting for a bus or trolley relatively enjoyable, even in the rain. There are several big, ugly parking garages, but fields of surface parking are rare. The city has filled in over the years, using parking lots for building sites or parks. Most downtown streets are very walkable and bikeable, at least west of the Willamette River. There are few negative impacts from freeways and ramps, although the massive bridges over the river dominate some areas. The downtown is compact and imageable with well-defined nodes, paths, and edges.

Baltimore: A (10)

Downtown Baltimore is served by both a subway (Metro) line and a light rail line. The Metro line connects the CBD with low-income neighborhoods to the east and west while the light rail system links major downtown attractions

Light rail in downtown Portland

such as Camden Yards (baseball) and the art museum. Both systems have stops only a few blocks away from the Inner Harbor. The harbor itself is served by a variety of water taxis and harbor shuttles that take riders to destinations such as Fells Point. In addition, there are two railroad stations facilitating trips from nearby cities, especially Washington, to the Baltimore waterfront and sports complexes. South of downtown, however, a tangle of freeways makes for a bleak environment, and just to the east of downtown, Highway 83 cuts off the urban core from the neighborhoods beyond. There are relatively few vast parking lots in the compact and walkable core.

Atlanta: A− (9)

Atlanta has one of the best rail transportation systems in any middle-sized metropolitan area in the country. The east-west and north-south MARTA lines meet at Five Points, the traditional center of the CBD, and connect nearly all of the downtown (and Midtown) subdistricts. MARTA is a subway in the urban core and above ground beyond. It also connects downtown and the busy international airport. There is lots of transit-oriented development and there are good bus connections at major stops. The system is well used at all hours, and the transit information is readily available. In some ways, however, the system

may work too well, in that there is intensive development around MARTA stations but activity falls off rapidly only a few blocks away. One can ride MARTA to a football or basketball game and then to dinner without ever going outside. The fact that people can travel easily below ground can make the downtown streets seem empty at night. Still, it is an excellent system. Nevertheless, in spite of good mass transit, there are lots of surface parking lots and garages in downtown Atlanta, and massive highway interchanges carve up the center city and dominate much of the landscape.

Denver: A− (9)

Downtown Denver is served by both light rail and a very successful downtown bus system. Buses converge at a downtown station and move up and down a bus and pedestrian mall along Sixteenth Street. Movement is easy along a clear corridor that stretches from the state capitol at one end to LoDo and Coors Field at the other. People can walk, take the bus, or do some combination of the two. Light rail carries people in and out of the CBD. The freeways are located well away from the downtown core and do not dominate the landscape. There is, however, still a lot of surface parking.

Sixteenth Street bus mall in downtown Denver

Cleveland: A− (9)

Downtown Cleveland is served by several rail systems. The Terminal Tower is so named because it is located over a rail terminal, and commuter lines (including one to the airport) still converge there. Light rail lines have also been constructed to connect the CBD core with the waterfront museums and the Flats entertainment district. There is also an enclosed walkway that connects the Tower City shopping mall in the Terminal Tower to Jacobs Field and the Gund Arena. Still, there are many ragged surface parking lots, especially on the eastern margins of downtown, and many streets offer poor walking conditions. Although freeways dominate several downtown entryways, this is somewhat balanced by the existence of spectacular and historic bridges over the Cuyahoga River. There are relatively poor connections to the museum and theater district at University Circle several miles away.

San Diego: A− (9)

Downtown San Diego is well served by a light rail system that brings three lines together in the CBD. One line goes to southern suburbs and Tijuana, and is used not only by commuters but also by tourists and by residents of Mexico. Lines also lead to the east and north. The downtown train station offers several trips a day to Los Angeles and Orange Counties as well as to northern suburbs. There is also ferry service to Coronado, and water taxis link several harborside hotels. A motorized "trolley" loops through the central city, Coronado, and the zoo and museums in Balboa Park. City blocks are short and reasonably walkable. Nonetheless, there are many surface parking lots, and a freeway interchange separates downtown from the attractions of Balboa Park.

Seattle: B+ (8)

There is an underground bus tunnel beneath the streets of downtown Seattle, but as yet there is no light rail. Congestion is a big problem in the city in that lakes separate many suburban areas from the urban core and bridges are overloaded. The old (1962) monorail system takes riders from the retail core to the attractions at Seattle Center but it covers less than a mile. There are plans for light rail, but it could just displace buses from the tunnel. Many commuters do come in from the islands by ferry each day, though many of these carry cars and so add to the congestion. The aesthetically problematic and noisy elevated Alaskan Way Viaduct dominates the waterfront. On the other side of the CBD,

however, Interstate 5 has been partly covered over by the tree-filled Freeway Park, thus enhancing connections with nearby neighborhoods. An addition to the railroad station is being completed and will facilitate visits to the sports complexes and Pioneer Square. Free bus transit in the downtown has encouraged people to explore the central city, but steep slopes sometimes make walking difficult.

St. Louis: B+ (8)

Downtown St. Louis is served by one light rail line that leads to the western suburbs, and it also connects several downtown attractions including Union Station, Kiel Arena (basketball and hockey), Busch Stadium (baseball), and the waterfront. Free lunch-time buses encourage downtown exploration. On the other hand, several wide roads leading out of the CBD make it easy to see a game or shop and then leave town quickly. A freeway separates downtown from the Gateway Arch, and a freeway also forms a barely penetrable wall south of the CBD. Ramps and entryways are awkward. There is lots of surface parking and a rather ugly garage at the waterfront. The bridges over the Mississippi, however, are spectacular.

Pittsburgh: B (7)

Light rail enters the edge of downtown Pittsburgh but does not link its parts together very well. Two incline railways on the south side of the Monongahela River add an interesting way to commute into the CBD. Boats shuttle people to ball games and other attractions, at least during the warmer months. Still, topography makes for a lot of congestion, ugly freeway ramps, and few optional routes. Highways dominate the riverfront and often cut off the city from the water, although some new parks and promenades are being created to soften their impact. Some interesting and historic bridges add character to the downtown. In general, the CBD is compact and quite walkable, although awkward street grids make bus routes difficult.

Minneapolis: B− (6)

Downtown Minneapolis focuses on the Nicollet Mall, a bus-only transportation corridor that ties the linear CBD together. The downtown is very compact, in part because of the second-level walkway (skyway) system, since most major buildings are tied into it. The combination of bus transit and the skyway system makes it easy to get around the core in all kinds of weather. As

yet there is no rail system, although plans are being made. On the west side of the CBD, freeways flow into a massive complex of mid-rise garages that wall off the city in that direction. They are functional but hardly aesthetic. A highway interchange cuts off the CBD from the art museums and residential areas to the south around Loring Park.

Providence: B− (6)

Downtown Providence is so small that it is easy to walk from one end to the other, although nearby residential neighborhoods are located on hills above the city core. Because of the major bus center in the plaza in front of City Hall it is easy to travel to surrounding neighborhoods. There is also a train station (Amtrak) between the downtown core and the state capitol, which enables visitors from as far away as Boston to visit regularly for shopping or cultural events. Within the downtown, where streets are narrow and curved, bus transit is not always easy. Walkways along the newly revitalized river, however, are inviting.

San Antonio: B− (6)

San Antonio has an award-winning bus system with good coverage. There is little sense, however, of a consensus transit node. The downtown streets are narrow with lots of awkward angles, and the giant superblock that includes Hemisfair Plaza and the convention center makes some connections and linkages difficult. Wide and sometimes elevated freeways act as unpleasant edges for the downtown on the east and west sides, and both surface parking and some rather intrusive garages dominate many parts of the downtown. A tourist "trolley" system and, of course, boats along the Riverwalk add charming options for intra-downtown movement. There is talk of a light rail system but none exists for now.

Charlotte: B− (6)

Although there is discussion of a future light rail system, for now Charlotte relies on cars and buses. There is a big bus terminal in the center of the CBD core close to the tallest landmark buildings, and bus options are easy to find. Even though an inner belt of highways separates the downtown from surrounding neighborhoods, large, straight streets make commuting in and out of the core fairly straightforward. There is a great deal of surface parking that tends to dilute the sense of urbanity.

Columbus: C+ (5)

Downtown Columbus has a semi–bus mall running the length of High Street, the mile-long "main drag," and this connects the Brewery District to the south with the Arena District to the north. Capitol Square and the office core are right in the middle. The bus system is adequate, but there is no rail system and the large size of the downtown makes walking from one end to the other problematic, especially in areas with lots of parking. The major streets are straight and easy to navigate, but there are large gaps in the urban fabric. The inner belt of highways that surround downtown is below grade and relatively easy to cross in most places. Massive parking garages are evident at several locations.

Indianapolis: C+ (5)

Downtown Indianapolis is characterized by having wide, straight streets and lots of handy on-street parking as well as some large peripheral parking garages. There is a good bus system, with fairly frequent buses converging on the core. There is a great deal of surface parking on every downtown edge especially around the university and south of the tracks near Eli Lilly Corporate Headquarters. The highway inner belt that borders the downtown core on three sides is far removed from the compact core, but an elevated rail line cuts through the heart of the CBD.

Phoenix: C+ (5)

Much of downtown Phoenix is quintessentially suburban, in that it is characterized by wide and busy highways with little on-street parking. One can move but not stop. The street grid is clear except where it is interrupted by superblocks and railroad tracks. A free shuttle operates within the CBD and there is a central terminal for city buses. Freeways are located well away from the CBD core, and a covered "Freeway Park" exists along the northern Central Avenue spine. Much of the downtown edge is made up of surface parking, although there are large garages adjacent to major attractions such as sports venues. Many "downtown" landmarks are difficult to walk to from the core, including the state capitol and art museum.

Summary: Transit Options

Portland 10
Baltimore 10

Atlanta 9
Denver 9
Cleveland 9
San Diego 9
Seattle 8
St. Louis 8
Pittsburgh 7
Minneapolis 6
Providence 6
San Antonio 6
Charlotte 6
Columbus 5
Indianapolis 5
Phoenix 5

Ranking Downtowns: Toward a Model of Spatial Organization

Like Italian hill towns or Mexican plazas, American downtowns all have a certain shared look or overall ambience yet have retained their individual personalities as well. Critics may say that all downtowns look alike, while boosters may argue that each one is uniquely attractive. The reality, of course, is somewhere in between. In this section, I first point out the particular attributes and problems found in specific cities and then generalize about downtown spatial structure. The goal is to clarify what works and what does not as cities enter the twenty-first century. I hope that smaller cities that are not as far along will be able to learn from the mistakes and successes of my case studies. In spite of having a great deal in common, downtowns still tend to be very different places, in part due to physical setting and historical accident and in part due to specific policies and programs enacted over the years. Finally, building on earlier work, I propose a series of generalizations on downtown morphology that I hope will elucidate the processes at work in the new American downtown. Just what is the American downtown all about? What myths should we discard as we develop new types of places and new attitudes toward them? How do we begin to think about what is happening to our central cities? I hope to contribute to new ways of thinking about American downtowns.

Table 6.1. Downtown Rating Table

City (Total)	Physical Site	Street Morph.	Civic Space	Office Towers	Retail Options	Hotel Conv.	Major Attrac.	Historic Dist.	Residential	Transit
Seattle (91)	10	7	6	10	10	10	10	10	10	8
Portland (87)	10	10	9	6	9	7	7	9	10	10
Baltimore (85)	10	6	8	7	8	9	10	9	8	10
San Diego (83)	10	9	5	6	9	10	6	9	10	9
Denver (81)	6	7	10	8	6	9	10	8	8	9
Cleveland (76)	8	5	9	10	8	6	9	9	3	9
Pittsburgh (75)	10	5	7	10	9	5	8	9	5	7
St. Louis (75)	8	7	10	7	7	8	9	7	4	8
Minneapolis (74)	7	7	6	9	8	7	8	7	9	6
Indianapolis (74)	6	9	10	7	9	7	9	6	6	5
Atlanta (72)	4	4	7	10	7	10	9	6	6	9
Columbus (72)	7	8	10	8	8	6	6	7	6	5
San Antonio (70)	9	5	7	4	9	10	7	8	5	6
Providence (63)	7	6	8	4	8	4	4	9	7	6
Phoenix (48)	3	6	5	5	3	4	9	4	4	5
Charlotte (48)	4	6	4	7	3	5	5	3	5	6
Possible Score (100)										
Possible points per variable (160)	119	107	121	118	121	117	126	120	107	118

Downtown Summaries: The Search for the Common and the Unique

Seattle: (91)

Downtown Seattle ranks first on the list as a place that has everything—busy waterfront, office towers, shopping, housing, and major attractions. It ranks lowest in street morphology due to several mismatched grids, no consensus "great streets," and some awkward and steep connections. It also lacks many truly grand civic spaces, although progress has been made with the creation of Westlake Park. Like many downtowns, however, central Seattle has been stretched into what may be an excessively linear arrangement. Squeezed between Interstate 5 to the east and the waterfront, downtown Seattle is about two and a half miles long and one-half mile wide. It is a long way from the Space Needle in the north to Safeco Field in the south. This linear pattern dilutes the sense of being in a place, especially since there is no real consensus center. However, the linkages are good, since the streets are generally walkable with lots of activity and amusement along the way. Bus routes and the monorail also help to tie things together, but downtown should really be a place that is tied together by pedestrian flows. One suggested improvement might be to create a grand civic plaza with a "famous" city hall and courthouse to reinforce the sense of city center.

Portland: (87)

In second place, downtown Portland is a perfect example of a good downtown in many ways. Nestled between the West Portland Hills and the Willamette River, downtown Portland occupies a pleasant and cozy site. The street grid is ideal, with small blocks and transit malls. The civic space is also good, with the statue of Portlandia and Pioneer Courthouse Square, giving the downtown a strong identity and sense of center. Even some of the relatively low scores, such as for office towers, may actually indicate successful renovation of older buildings rather than the inability to create quality office space. The main problem with downtown Portland is that, in part due to its compact size and tight boundaries and political pressure from the east side communities, many activities, such as the convention center, sports arena, and several hotels, have been built on the other side of the wide Willamette River in a modern and unremarkable secondary downtown. While not an altogether bad idea, this sepa-

Westlake Park, a new civic plaza in downtown
Seattle

ration of functions dilutes the importance of the downtown core and diminishes its potential to be a truly great center, even with excellent trolley connections. The space that is becoming available in the once industrial Pearl District
is being developed primarily for housing. A major suggestion might be to improve the quality of the secondary downtown by attempting to replicate some
of the best qualities of the old core. If possible, broad highways and freeway
ramps could be softened and better integrated with a traditional grid of tree-
lined streets.

Baltimore: (85)

Baltimore's high ranking is mainly a result of its revitalized Inner Harbor
rather than the successful development of its traditional downtown core.
Downtown Baltimore has become the quintessential "city as theme park" for

the greater Baltimore-Washington urban region. As the center of activity has moved to the harbor, however, the downtown core has suffered by comparison. As the urban renewal projects of the 1950s such as Charles Center become more dated and sterile, people vote with their feet and head for the water. This is especially true as major attractions are being developed ever further outward from Fells Point to Federal Hill. To counter this trend, a cultural district is being strengthened on the northern edge of downtown, but physical and perceptual linkages between all the parts are a bit weak. On the other hand, central Baltimore has a lot to offer and may well epitomize the downtown of the future—a series of separate "worlds" à la Disneyland, each with its own theme and identity. The core civic and financial area may play second fiddle to entertainment around the edges. The diversity of downtown is enhanced by the presence of gentrifying neighborhoods close to low-income public housing projects, and this urban diversity is part of the sense of place. A suggestion might be to make every effort to return Charles Street to its former status as a consensus "great street" and create clearer linkages to the surrounding districts.

San Diego: (83)

Downtown San Diego lacks many of the characteristics of a traditional, strong city core, but it is doing a good job with the things that it has chosen to emphasize. There are few headquarters office towers or grand civic spaces, and most of the major attractions from sports facilities to museums are located elsewhere. In recent years, however, the downtown has become a center for recreation and residences. Seaport Village, the Gaslamp Quarter, harbor excursions, hotel and convention facilities, and the whimsical, postmodern Horton Plaza shopping center bring people into the center of town. Downtown is nearly square in shape, with every destination reasonably close to most others. Already about twenty thousand people live downtown, and at least three major areas have been designated as primarily residential, with dozens of new housing projects under way. Given the overheated housing market in southern California, however, very few of the new units are affordable, and considerable gentrification is going on. A major problem downtown has been the stalled baseball park project—a vast empty field of dreams for several years. In addition to finishing that project and filling in around it with sympathetic development, a major suggestion might be to improve connections with nearby Balboa Park, where many of the museums (as well as the zoo) are located. At present, the park seems distant because it is cut off by a freeway and because

pedestrian access is inadequate. Perhaps some kind of "freeway park" à la Seattle could improve things.

Denver: (81)

Already ranking among the better downtowns, Denver has the potential to move up on the list. Not only could downtown be greatly enlarged by expanding into the empty field that separates it from the South Platte River, but its physical site could be enhanced in the process by integrating the river into the overall sense of place. At present, the area is being gradually filled in with attractions such as Elitch Gardens Amusement Park and the Pepsi Center arena, but with careful planning, the downtown grid could be expanded into the district. At present, the vast empty spaces only serve to give downtown a sense of disconnection to west-side neighborhoods even as links to neighborhoods in other directions are improving. While downtown Denver already has lots of office space, civic space, and major attractions, it ranks low as a shopping destination. Although there are plenty of specialty shops and galleries, there are no major anchor stores. This could easily be fixed, but it would probably mean building a major downtown mall like those in Indianapolis or Columbus. There is plenty of space and Denver could eventually double the size of its

Downtown Denver has plenty of room to expand toward the South Platte River.

downtown, but this must be done carefully or the sense of a compact and legible center could be lost. Whereas tight boundaries constrain Portland, their absence may be a mixed blessing for Denver.

Cleveland: (76)

And now for something completely different. Downtown Cleveland garners points as the gleaming core of a still troubled city. While most of the cities discussed above have relatively strong and desirable supporting neighborhoods around the downtown, abandoned ghetto housing and obsolete industrial districts dominate Cleveland's "inner city." Things drop off drastically since the core lacks nearby support districts for motels, restaurants, and residences. Even getting to University Circle, the home of the art museum and symphony, requires a trip through several miles of decaying landscape. Still, downtown has a lot to offer. Office towers, retail malls, historic entertainment districts, theaters, and museums all make downtown Cleveland vibrant, but residential areas have been slow to develop, in part because it is impossible for them to link up with existing attractive neighborhoods. New lakefront projects could help to establish a greater sense of amenity, but more is needed. The main suggestion today might be to work on developing surrounding neighborhoods, especially on the east side but also in the Cuyahoga River area and Ohio City to the west. The goal would be to make downtown Cleveland a more integral part of a better city.

Pittsburgh: (75)

Downtown Pittsburgh has a stunning but awkward physical site that gives it both a powerful sense of place and the inability to easily expand. The CBD is very compact, with narrow streets and a sense of urbanity, but most space-extensive activities must be located elsewhere, increasingly across the rivers, especially since "The Hill" to the east is now largely rebuilt with housing. While crossing a bridge in a vehicle can be a wonderful kinetic experience, walking can be unpleasant on a cold and windy day. The challenge will be to better integrate the areas across the rivers that are now gradually being incorporated into downtown and to improve the linkages to nearby neighborhoods. Greater use of boats, enclosed walkways, trolleys, and shuttles will be needed to diminish the impact of topographical and psychological barriers between the CBD core and the sports and museum districts beyond. As in Cleveland and St. Louis, much of the "art and culture" activities are located in a separate dis-

trict, but perhaps an increasing number of them could find homes in new river-front complexes or perhaps in the deindustrializing "Strip" district.

St. Louis: (75)

Downtown St. Louis is also an island in the midst of a struggling city. Both the central parts of the city of St. Louis and the city of East St. Louis across the Mississippi River act as liabilities to the revitalizing core. Acres of empty fields make downtown seem isolated, merely a destination for special events rather than a desirable place to live and work. Decades of clearance programs, public housing disasters, and deindustrialization have taken a toll on the central city. But the downtown has tremendous potential. The river provides a major amenity, and grand civic spaces abound. There are a variety of major attractions and a plentiful stock of historic structures. Obviously, a major task ahead is to revitalize the neighborhoods around downtown, and this is already being done on a small scale, since lower-density and pleasantly designed townhouses have replaced some of the more notorious projects such as Pruitt-Igoe. Because most of the housing is still subsidized, the city does not have the kinds of afford-ability issues facing San Diego and Seattle. Getting more people to live down-town may depend in part on strengthening the reality and the image of a con-

The potential is there for a huge loft district in central St. Louis.

tinuous sector of activity from the CBD core to Union Station and on to the West End and Forest Park. At present, there are too many gaps. Another suggestion, and one that is being followed already, is to find uses for the abundant historic architectural stock that is now generally underutilized downtown.

Minneapolis: (74)

Downtown Minneapolis has all the pieces, but somehow it does not work quite as well as it should. The downtown is very compact due to the perceived desirability of a link to the skyway (elevated walkway) system. The walkways make for a vibrant indoor city but are aesthetically problematic outside, in that they block views down the street and spoil architectural perspectives. They also can give the impression that the street or sidewalk is not the real city. Big garages and freeway ramps form a wall to the west but are functional in that they keep excessive commuter traffic out of the CBD core. Nicollet "Mall" is the consensus main street, and there is a good sense of center. The river to the north and the lakes to the south provide amenities and a strong sense of place for downtown and have recently attracted a large number of residential projects, though both are relatively poorly integrated into the CBD. The huge University of Minnesota is not far from the downtown core yet seems to have surprisingly little impact. Town-gown linkages could be enhanced. Major suggestions might be to better integrate the river and the university into the downtown. Downtown could also use a centrally located, grand civic space of some kind.

Indianapolis: (74)

While all of the downtowns on the list have made great strides over the past two decades, Indianapolis is a leading candidate for the "most improved" award. Aside from having the state capitol, downtown had few attractions as recently as the 1970s. Today, the city has a lively, compact core and a strong image as a sports and convention center. A downtown shopping mall, complete with theaters and restaurants, has helped to spur retail in surrounding blocks. The downtown core still has relatively poor connections to most of the surrounding neighborhoods, although major improvements are being made, especially toward the White River. An elevated railroad and vast fields of parking isolate the core from neighborhoods to the south, and the once vibrant northern spine has deteriorated as the core has imploded. Large facilities for sports, while helping to give the city an image, sometimes act as walls around the

Indianapolis has removed railroad tracks and reconnected with the White River via a park and canal.

downtown core. A major improvement would be to gradually fill in the spaces between downtown and the neighborhoods beyond with housing and downtown support districts. This is already happening, but it is a slow process.

Atlanta: (72)

Downtown Atlanta suffers from the lack of a consensus core since there is no significant physical site or civic center to give the place a sense of coherence. Once again, most of the pieces are there but they do not quite fit together. The state capitol is peripheral to the downtown core and is largely surrounded by a freeway interchange. The new Centennial Olympic Park has tremendous potential to be a civic center, but the buildings nearby are massive and fortresslike and give the place a cold, remote feeling. Many of the major "downtown" attractions, such as the High Art Museum and most theaters, have moved northward into Midtown, accentuating a prevailing perception of a white "north" and black "south" downtown. Underground Atlanta struggles to remain the only major human-scale architectural attraction downtown. A convoluted street system also makes downtown seem less coherent than most. As activities have moved northward along the Peachtree spine, the city seems too much like

a linear strip with rapid fall-off around the edges. Still, stunning office towers and massive convention and sports venues make the city a major destination. To a very real degree, however, the buildings are too big. The downtown is made up of huge but unrelated megastructures. The main suggestion would be to create as many human-scale streetscapes as possible in and around the core in order to mitigate the current dominance of enclosed, self-contained spaces.

Columbus: (72)

Downtown Columbus has perhaps the best consensus central square in the study, but there are still not quite enough major attractions to make the CBD a significant destination. The recent construction of the Nationwide (NHL) Arena has spurred the development of dining and entertainment in the emerging "Arena District," but that could just pull activity to the edge of downtown. A newly expanded science museum could also have some impact in revitalizing the moribund area west of the Scioto River. The normal activities of business and shopping are well developed and in nice settings, but they could use some beefing up, as many gaps in the urban fabric remain unfilled. Also, the subcores and historic districts at the northern and southern ends of downtown

Downtown Atlanta is dominated by a number of very large and often poorly linked megastructures.

are separated from each other and from Capitol Square by fairly long distances. Very few people live in the heart of the city, though nearby historic neighborhoods are attractive. In many ways, Columbus is the polar opposite of Cleveland, which has a very strong core but weak surrounding neighborhoods. Filling in downtown parking lots with mid-rise housing, office, shopping, and entertainment destinations would give downtown a bit more bulk and coherence. Given the distances between major attractions, a light rail system would also help.

San Antonio: (70)

There is little doubt that downtown San Antonio is a wonderfully attractive place that visitors enjoy immensely. It has more hotel rooms than any other downtown on the list and probably more restaurants and gift shops. The main weaknesses revolve around the relative lack of nontourist downtown functions. There is relatively little office space, with only a handful of moderate-sized buildings on the edge of the core. The downtown functions as much as an Alamo theme park as an employment and decision-making center. The CBD also seems a little disorganized and poorly connected due to the combination of the below-grade meandering Riverwalk and the lack of a clear grid. In spite of a very attractive historic residential area nearby (King William), there are relatively poor connections to most surrounding neighborhoods. Revitalization efforts are under way on older commercial streets, however, and the city may act as a role model for creating a strong downtown with relatively few traditional central business district functions.

Providence: (63)

Downtown Providence is unusual in that it is much smaller in area than any of the other downtowns under consideration. Its compact size is a function of both its age and the fact that it is nestled in a restrictive physical site, surrounded by hills and water (as well as a freeway). There is very little room for expansion, although a new mall has been constructed between the CBD core and the hill-top state capitol. There is danger, however, that its success will further marginalize shopping in the old core. Still, the streets are narrow and everything is close together. The city lies in the shadow of Boston and is not a major central place and has few major attractions. It has been growing very slowly for a long time, and in some ways shows its age. Many older downtown buildings remain but they often house art galleries or cafes and there is relatively little

Phoenix could use more human-scale settings such as the Arizona Center.

growth in high-level office space. With Brown University and the new mall, the downtown could become a retail and cultural destination for several nearby New England communities. Downtown must concentrate on the things it does best and try to make good use of its historic setting and architecture.

Phoenix: (48)

Everything in downtown Phoenix is too big and/or spread out. The streets are too wide, the activities (especially the state capitol, the library, and the art museums) are too distant, many of the blocks and superblocks are too large, and the massive convention and sports facilities located in the core further separate potential interaction zones. An attempt has been made to identify and publicize a smaller version of the downtown that includes only the core, and this is a good idea. Copper Square, as it is called, is small enough to be walkable but still contains a large number of big-footprint structures such as the convention center, parking garages, the baseball park, and the sports arena. The Arizona Center provides a nice setting for entertainment with cafes, trees, and water ponds in the (afternoon) shadow of two office towers on the edge of the square. Without small streets and buildings and with no physical feature to fo-

cus activities, the downtown seems as much like an edge city office park as a traditional city center. There are few residential neighborhoods nearby aside from the few towers on the North Central Avenue spine and in the Roosevelt Historic District, but this is gradually changing. Like Atlanta, downtown Phoenix needs more human-scale architecture.

Charlotte: (48)

Charlotte is a city on the make, and this is both good and bad. It has experienced a boom in recent years but in the process it has destroyed much of its older character. As a southern city, Charlotte never had much in the way of ethnic districts or downtown industrial zones that could help to provide a sense of place and history. Since the metropolitan area was very small as recently as the 1950s, the "big league" downtown has had to be built largely from scratch, and so it seems like a gleaming new place but with few nooks and crannies for supporting activities. Office towers have appeared but retail and residential uses have been slow to follow. While the site is not unattractive, there are no special water features or topographical elements to focus the downtown core. The football stadium brings people into the area, but there are few symbiotic connections with the downtown. Charlotte simply needs more "stuff" to fill in around the office core—especially shopping, entertainment, and housing. The grid is clearly organized and could be quite walkable with more interesting destinations along the way.

Major Themes in Downtown Spatial Organization

In reviewing these individual case studies, several themes or trends appear regularly. The most notable are: (1) the trend toward linear downtowns focusing on one major "banner" street at the expense of once valued support streets; (2) the identification of separate "worlds" or theme districts within the downtown, each emphasizing its own look, function, and diurnal or seasonal time slot; (3) the enhancement and even creation of physical amenities that can serve as anchors and image-makers for downtown projects; (4) the trend toward big-footprint buildings that have transformed the downtown from a setting where there are spaces between buildings to one characterized by buildings lost in space; (5) the increasing competition between privatized spaces and revitalized civic spaces to act as the "real" consensus centers of activity; (6) a slowdown in office construction and the continuing loss of industrial and warehouse functions coupled with the search for new space-extensive land uses, especially res-

idential and recreational, to take up the slack; and (7) a greatly diminished concern with classic "urban" problems such as crime, congestion, and pollution coupled with increasing efforts to make downtown a family destination with lots of special events and big crowds.

The Linear Downtown

Linear downtowns go back a long way in American urban history. New York City's CBD grew out along several miles of the Broadway–Fifth Avenue spine with nearby Third or Seventh Avenues being seen as peripheral. In Chicago, the downtown followed the State Street–Michigan Avenue axis, often with sharp drops in land use intensity on either side. Many of the middle-sized cities in this study have emulated those patterns in recent years. Perhaps Atlanta affords the best example. Everyone wants to be on Peachtree Street. Indeed, north of the CBD core, there are two Peachtree streets, one for each direction, and in the core, street names have been changed so as to have the word Peachtree included (such as Peachtree Center Avenue). Greater downtown Atlanta now extends for nearly four miles along the Peachtree ridge from the southern inner belt to the museums and theaters in Midtown. At the same time, many streets only a few blocks off Peachtree remain empty or underutilized. Peachtree has followed the Broadway–Miracle Mile American tradition. Few other downtowns are quite so purposefully linear, but some come close.

Downtown Seattle is becoming more linear by the week as new developments are concentrated around the new sports complexes and train station to the south and Belltown and the Seattle Center to the north. Squeezed between the water on the west and Interstate 5 and Capitol Hill on the east, downtown Seattle follows Fourth Street (with several parallel helpers) for over three miles from Safeco Field to the Key Arena. While there is no one "great street" as there is in Atlanta, everything is a couple of blocks from the main spine.

Downtown Portland extends for a mere mile and a half, not counting the "east of the river" extension, but it is still developing in an increasingly linear pattern. As the Pearl District to the north is gradually incorporated into downtown, the north-south extensions will seem more pronounced. The combination of a highly successful bus mall and a linear strip of green parks reinforce the linear dimension.

Downtown Denver has also become increasingly linear with the success of the Sixteenth Street Mall—a combination pedestrian and bus mall lined with

Atlanta, like many American downtowns, features a dominant linear spine of major towers.

benches, cafes, and entertainment venues. The mall extends from the state capitol to LoDo, with office towers in the blocks to the northeast and hotels, a convention center, and a performing arts center to the southwest. Most of the
"movement" in the core follows this dominant spine.

Downtown Columbus occupies a rectangular area with a classic PLVI at Broad
and High Streets. While Broad Street has offered some fierce competition from
time to time, High Street has always been the main commercial spine. Recently,
major activity centers such as the Brewery District to the south and the Arena
District to the north have accentuated the existence of a dominant linear spine
of about one and a half miles long. Very little "core" activity is located more than
a block or two off High Street, and a special bus lane reinforces the connections.

In San Antonio, the generally east-west orientation of the river (and Riverwalk) defines the downtown. No one wants to be very far from the river, and

the major commercial streets run parallel to it. Although the linear pattern is less obvious here because the Riverwalk is below grade and somewhat convoluted, the east-west flow of activity is dominant.

In Minneapolis, the Nicollet Mall running from (close to) the Mississippi River on the north to Loring Park on the south creates a dominant spine that is reinforced by important helper streets such as Hennepin and Second Avenues. Most everything from parades to street markets takes place on these north-south streets. Charlotte is an even better example of a one-street town, with nearly everything of importance concentrated on Tryon.

Some downtowns have been very linear historically but are now taking steps to diminish the dominance of the major spine and spread activity to other streets. Downtown Baltimore, for example, is still focused on the north-south Charles Street spine that runs from the convention center to the Mount Vernon cultural district. The increasing importance of the waterfront and Camden Yards, however, has moved much of the action to the east and west along Pratt. Similarly, downtown Indianapolis has always had a north-south spine running along Meridian Street from Union Station in the south, through Monument Circle to Veteran's Plaza to the north. Recent developments to the east and west have accentuated the importance of Washington Street. Downtown Phoenix has long consisted of little more than a northern spine along Central Avenue, but recent attempts to create a "Copper Square" have been aimed at concentrating downtown activity. Downtown Cleveland has also imploded as the core has been rediscovered at the expense of the Euclid Avenue spine. Downtown San Diego's spine was always Fifth Avenue running from the waterfront toward Balboa Park, although various attempts were made over the years to create competing east-west streets such as Market and then Broadway, especially since the flight path squelched enthusiasm for many locations on North Fifth.

A few downtowns, Providence and Pittsburgh for example, were never really able to achieve very much linearity in part due to site constraints. Pittsburgh is still trying, however, as nightlife is being extended eastward through the "Strip" district, and the Fifth Avenue retail corridor has been slated for controversial redevelopment.

I have previously attempted to generalize the linear character of the American downtown by creating a "model" of downtown structure. While the original model was largely based on larger cities such as New York, Boston, Philadelphia, and San Francisco, it works pretty well for many of the cities in this study. Nevertheless, a variation on the theme is included here as well.

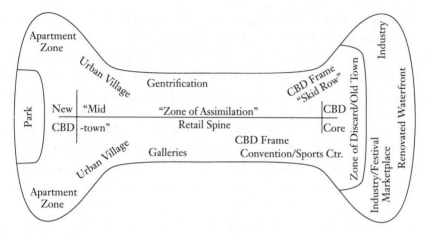

A Model of the American Downtown
As North American downtowns have grown and changed over the years, many have
assumed a predictable shape and size. The best model to describe them is a dumbbell
with the old waterfront/skid row/central business district at one end and a new up-
town/museum/nightlife district (and sometimes a park) at the other. The two ends
of the dumbbell are typically connected by a relatively thin spine focusing on a major
retailing street. Although there are many variations on the theme, the model de-
scribes the nearly universal process of CBD extension and reconcentration.
From Larry R. Ford, *Cities and Buildings: Skyscrapers, Skid Rows, and Suburbs*
(Baltimore, Md.: Johns Hopkins University Press, 1994), p. 86.

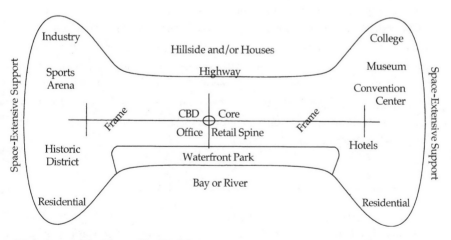

Variations on the Linear City Model
In this model, the downtown has refocused on a waterfront amenity. As deindustrialization along the waterfront has taken place, parks, housing, hotels, and historic districts have emerged.

These variations on the linear city model first presented in 1994 aim to recognize that many cities have remained focused on a traditional business and government core but have developed more space-extensive support nodes further out, usually along a dominant spine. Big-footprint buildings such as convention centers and sports arenas cannot easily be accommodated in the heart of the CBD because of land assembly problems. Neither is it wise to have massive structures in the center of the city, since they usually require too many blank-walled superblocks and thus spread out and dilute urban life. These big structures are used to soak up excess space on the downtown periphery that was once used for such things as warehouses, prisons, factories, and marginal housing. While they sometimes foster lively entertainment districts nearby, the need for big parking garages and storage areas usually results in some dead zones as well. The center city thus remains dense, vertical, and compact while the ends of the dumbbell are occupied by large, horizontal structures.

In several cities, massive government structures and associated civic space occupy settings above and away from the business core. The best examples are state capitols such as those in Denver, Atlanta, or Providence, but courthouses, libraries, city halls, and federal complexes can anchor nodes quite apart from the world of commerce. Often such separation was deliberately planned in the past in order to keep government free from the grime and possible corruption of the downtown.

The new version of the model also assumes that the linear downtown can run parallel to the waterfront as well as away from it. As deindustrialization has occurred and as clean-up measures have been instituted, many downtowns have sought to hug their rivers or harbors rather than run from them, as was the case in classic zone of discard/zone of assimilation models. The activities and functions of the linear downtown can thus be enhanced and reinforced by access to real or potential waterfront amenities. Such locations are ideal settings for new hotels and housing projects.

Downtowns as Separate Worlds or Theme Districts

American downtowns are no longer divided into simply core CBDs or central business districts surrounded by supporting "zones in transition." Nearly every downtown of any size has now been divided into between five and ten districts, each with its own unique identity and appeal. Characteristics that were once thought of as inferior, such as older, somewhat risqué quarters or in-

dustrial and warehouse zones, are now seen as contributing to the variety and complexity of the downtown experience. The better the downtown, the more different districts and identifiable personalities there are.

Seattle, for example, has Pioneer Square, the International District, the waterfront (with several subdistricts), Belltown, Westlake Park (retail), Seattle Center, the financial district, and the new ballpark district, just to name a few. Baltimore's downtown includes a variety of places from the museums and shops of the Inner Harbor to the Mount Vernon cultural district. Columbus has a new Arena District and San Diego has just completed extensive renovations and new signage for Little Italy. The once industrial Pearl District in Portland is nearly as big as the downtown core and is being completely renovated, largely for residential purposes. The list goes on.

Downtowns have followed Disneyland in providing a variety of separate worlds and experiences. The old attractions of business buildings, department stores, and major hotels are not enough anymore. Even the convention trade is attracted to cities with a variety of "places to go" within walking distance or a short cab ride from the meetings. Within a few blocks of the massive San Antonio Convention Center, a visitor can find La Villita, the Riverwalk, El Mercado, and the King William Historic District.

The trick is to have just enough, but not too much, space (as Disney discovered long ago). Downtown Pittsburgh's Golden Triangle is too small to have enough variety, but efforts are being made to better integrate settings on the other sides of the rivers. Downtown Phoenix, on the other hand, covers too much territory and some of the attractions, such as the new library or state capitol, are so distant that they may not be considered part of downtown. How big can downtowns be? How small? The designation of Copper Square is an attempt to limit downtown and create a consensus core.

I have developed a model of "Central City as Disneyland" in an earlier book, but it is still appropriate for understanding current trends in downtown structure. While each city has its own variations on the Disneyland model, most have essentially the same major components—financial districts, shopping districts, historic warehouse districts, sports districts, upscale residential nodes, ethnic villages, waterfront promenades and/or green parks, entertainment or "fun" zones, theater districts, museum and art gallery "cultural" districts, and hotel and convention center districts. Boat rides, observation decks, aquariums, and even zoos are also increasingly downtown attractions. Even though most downtowns have similar districts, they remain remarkably different.

Since most of the spatial arrangements have grown up over the centuries, rather than as part of planned theme parks, the pieces are seldom predictably organized, and this is as it should be.

The Enhancement of Physical Sites as Downtown Icons

Decades ago, cities were proud of their new buildings, boulevards, plazas, and train stations. These were the things that most often appeared in "post card" views of the downtown. Rivers and lakes, if featured at all, were often shown from a distance or as evidence of the industrial might of a particular place, complete with barges, grain elevators, and smokestacks. Today, physical features, especially water features, have taken over the foreground as part of the new recreation-filled downtown.[1] Parks and green spaces, often full of joggers or bicyclists, also rate as important downtown icons. As rivers, lakes, and harbors have been cleaned up and lined with green promenades, downtowns have not only moved toward them but, in many cases, wrapped themselves around them so as to maximize contact.

Of course, the downtowns with the most potential for physical feature enhancement often have the biggest obstacles to overcome. The Cuyahoga River meanders through downtown Cleveland on its way to Lake Erie, and triangular downtown Pittsburgh is surrounded on two sides by riverfront land. Not long ago, however, no sensible person would have gone near the water in either place. In Portland, the Willamette River, as well as the mighty Columbia, often flooded low-lying parts of the city, including much of the downtown. In downtown Providence, the rivers were covered over for highways and railroads, while downtown Indianapolis simply turned its back on the nearby White River. The core of downtown San Diego was so cut off from the bay fifty years ago that few downtown visitors ventured over to see it unless they were taking a ferry. In recent years, all this has changed.

Not only have downtowns sought to rehabilitate and reconnect with physical features, they have sometimes built them anew, or at least reinvented them. The San Antonio River is a largely artificial creation today, and Indianapolis is emulating its success with the creation of a canal walk, featuring a redesigned waterway along the western side of the CBD. Providence has created Water Place Park and Riverwalk as foreground for its new downtown shopping mall. Downtown Denver is lining parts of Cherry Creek with green walkways in order to create a connection with the small stream, and Cleveland is doing the

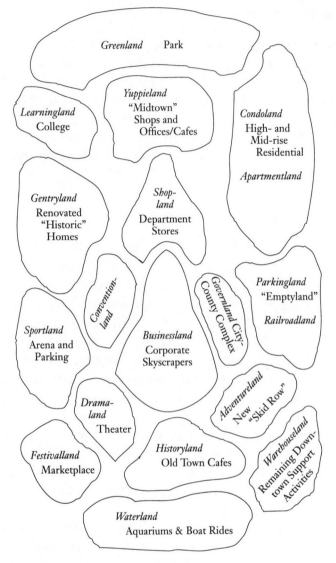

Central City as Disneyland

Over the past century, the typical North American downtown has been extremely dynamic, with some land uses expanding rapidly at the expense of others. Architectural changes have been equally noticeable, with continuous adjustments in building size, style, and function. Recently, a certain stability has begun to set in as particular areas have been given over more or less permanently for use as office towers, recreational waterfronts, convention and sports arenas, historic districts, residential districts, and the like. A combination of the large scale of new development and urban design ideologies has brought about a new kind of central city: a city of separate worlds.

From Larry R. Ford, *Cities and Buildings: Skyscrapers, Skid Rows, and Suburbs* (Baltimore, Md.: Johns Hopkins University Press, 1994), p. 91.

same with the Cuyahoga River. Of course, with extensive flood control on rivers and dredging and land filling in harbors, it is difficult to say what is natural in most settings.

For downtowns with little or nothing in the way of potential physical icons, such as Atlanta, Charlotte, or Phoenix, the best hope is to create a park with fountains and perhaps a pond as a foreground for the city skyline. Canyonlike business blocks and busy sidewalks filled with shoppers are not enough anymore.

Big-Footprint Buildings as Walls and Gates

Throughout most of American urban history, downtown lots were very small, even compared to those in many European cities where sizeable palaces and monasteries occupied space in the center of the city. In New York, Philadelphia, Boston, and other major cities, lots were often twenty-five feet wide and rarely more than fifty. The construction of big buildings required the long and sometimes cumbersome process of lot assembly. Although the massive urban renewal projects of the 1950s and 1960s made large spaces available, for a long time there was not much demand and vacant fields or parking lots dominated many downtown fringe areas. Infatuation with the modern look of skyscrapers in a park lessened the crisis, since it was quite acceptable to build a tower here and there separated by vast grass or concrete spaces. By the 1980s, however, the demand for huge buildings with immense footprints had arrived. Sports arenas, football stadiums, convention centers, parking garages, shopping malls, museums, and performing arts centers are all essentially horizontal structures that take up a lot of space.

Every downtown examined here has at least one such megastructure and most have several. Since the normal location for these buildings is on the periphery of the downtown core, their location has several implications for not only the internal structure of downtown but also for linkages with surrounding neighborhoods. Monster buildings can both channel activity and act as a wall, shutting off interaction with the outside world. In essence, they create both city walls and entry gates in shaping downtown morphology.

Downtown Atlanta is the best example of a place dominated by huge, fortresslike structures. The CBD has quite a few such buildings, especially around Peachtree Center to the north and the state capitol to the south. The biggest examples, however, have been wisely located over railroad tracks on the

Big-footprint buildings such as convention centers often form walls around the downtown core
as here in San Antonio.

edge of the CBD. In the process, an immense wall has been created. The downtown is "protected" from low-income, largely African American communities
on the south and east by massive freeway interchanges, but no highway was
built to the west. Today, the Georgia Dome, the Georgia World Congress Center, the Phillips Arena, and the CNN Complex create a buffer that makes the
neighborhoods beyond seem very distant from the downtown core. The same
effect has been accomplished, largely by accident but sometimes on purpose,
by such structures as the Alamodome in San Antonio, the Ericsson Stadium in
Charlotte, and the Civic Arena in Pittsburgh.

   In some cases, problematic residential neighborhoods were cleared long before the construction of such barriers, or in other cases, they were never there.
In Seattle, the sports complexes south of downtown separate the city from a

A massive convention center as a city wall in Atlanta

huge district of factories and warehouses and so help to deindustrialize the CBD psychologically. In Denver, the convention center and performing arts center megastructures separate the city core from what was once a low-income neighborhood but is now a university campus. Still, the wall effect is evident.

While they nearly always play some role in the creation of walls and barriers, megastructures can also help to channel activity so as to create necessary threshold populations for urban life. Such is the case in Columbus, Ohio. The convention center is just to the north of the CBD core along High Street, the city's main drag. Even though the Eisenmann-designed structure was built to look like a series of small geometric forms so as to diminish its boxy mass, the center has inevitably deadened the east side of the street. On the west side, however, a row of small shops and cafes has been rehabilitated to serve the needs of conventioneers. This strip encourages people to wander northward a block further to the Short North/Victorian Village area. The convention center and its related neighborhood thus form a tunnel between the downtown and the nearby historic district and help to channel visitors between the two. Similarly, the combination convention center and TWA Arena in St. Louis may help to direct people to the nearby Laclede's Landing Historic District.

Competition between Civic and Private Space

It is fashionable to say that American cities have abandoned the quest to build and maintain good civic space and that the real centers of our cities are shopping malls and office building atria. While there is some truth in this, it has been greatly exaggerated. In the downtowns examined here, good civic space not only still exists, it is usually being expanded and upgraded. While the shopping malls may have more people than the parks and courthouse squares, especially at night and during inclement weather, such places were never occupied twenty-four hours a day. Indeed, in most cities, long-term users would very likely have been arrested for vagrancy. Still, good civic spaces are usually well used at lunchtime and when there are special events. The poor (homeless?) still sit on park benches in today's cities, from Woodruff Park in Atlanta to Pioneer Square in Seattle, just as they have for the past century. If anything, these places are much better than they were thirty or forty years ago.

Both the total amount and the overall quality of civic space is better in some cities than in others, but a key question is the degree to which it is integrated with, versus segregated from, the new privatized spaces. Indianapolis has an abundance of civic space, but some of it is peripheral to the center of downtown activity. In Columbus, Capitol Square serves as the city center because it is connected directly to major office buildings, theaters, the shopping mall, and (almost) to the riverfront. We need a great deal more information about the flows to and from downtown public spaces to flesh out our concerns about their size and design. Decades ago, Whitney Seymour observed in *Small Urban Spaces* that tiny fragments of well-located space can work wonders in reinforcing busy commercial districts.[2] Subtle changes in location or access, however, can spoil the effect. On the other hand, large parks and plazas can form barriers to movement, especially at night and during blizzards and heat waves.

It is hard to argue these days that civic space alone should anchor urban centrality. From the agoras of ancient Greece to the gallerias of nineteenth-century Europe, covered shopping and dining have competed with purely civic spaces as major gathering places. In the age of air-conditioning and central heating, this is even truer. Multilevel downtown malls and food courts have added a great deal of "life" to downtowns, especially when they feature daily concerts or art exhibits. Civic and private spaces can, however, form a symbiotic relationship with them. Good civic space is often "on the way" between work, play, parking, and shopping. If people have to make a special effort to get there, they

probably will not go, because Americans are famous for being workaholics with little time for frivolous hanging out. "Accidental" gatherings work best.

The issue of who can use civic space, for what purposes, and at what time should be obvious. Political rallies, street entertainers, small vendors, and drowsy nappers should all be welcome in civic space, but herein lies the problem. When civic space is isolated from private space and off in a corner of downtown, marginal (poor) users are not usually viewed as a disruptive influence. No one sees them anyway. If, on the other hand, civic space and private space are integrally connected, disheveled visitors can not only be bad for business but can deter others who feel vulnerable from using the space. Right or wrong, homeless people sleeping in the doorways of shops and offices are not likely to be popular. Perhaps the trick is to achieve some intermediate degree of spatial integration that allows for a high level of centrality coupled with tolerant coexistence.

In the sketch below, civic space is surrounded on three sides by core activities with good, effortless connections to a central fountain. Toward the river, the space is more likely to be used purposefully by lunchtime strollers, joggers, lovers, and retirees and others just seeking to hang out. During special events, the space would be full of people, but most of time there is room for everyone. The model was inspired mostly by Portland and Pittsburgh but there are elements from several cities.

The Search for New Activities to Fill Vacated Space

During the heyday of urban renewal in the 1950s and 1960s, it was widely assumed that office space would be the main feature in most redevelopment projects. The gleaming new "Golden Triangle" in Pittsburgh, Erieview in Cleveland, and the Charles Center in Baltimore were all dominated by office towers. In recent years, especially as warehouses and railroad tracks have been removed, new types of districts have had to be created to fill the space. The demand for downtown office buildings has cooled in most cities, at least temporarily. Ballparks and convention centers have taken up significant amounts of space, but there are still some rough and underutilized edges around most mid-sized downtowns that need bolstering.

The main thrust has been to create residential neighborhoods or theme districts, complete with villagelike subcenters and parks. Some have resulted from historic preservation efforts and have used historic or current ethnic identity as the prevailing theme. Others are largely new. Sometimes historic features

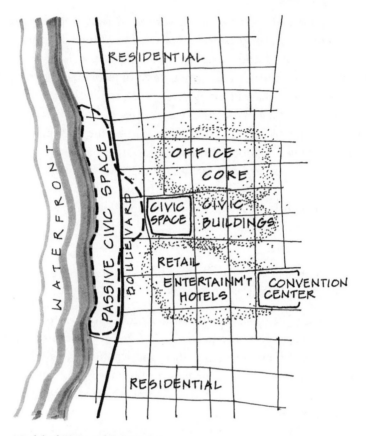

Model of Civic and Private Space

such as breweries, port facilities, and factories provide the theme for residential and recreational projects. Examples include German Village in Columbus, the Flats in Cleveland, Crawford Square in Pittsburgh, the Pearl District in Portland, and the Martin Luther King complex in Atlanta. In all cases, the result has been an expanded definition of "greater downtown" involving places that visitors and residents alike should explore.

In other cities, space-extensive activities not normally found in American downtowns, such as zoos (Indianapolis) and amusement parks (Denver), and university campuses (San Antonio) have been moved inward to bolster the central city. These may or may not be considered as truly downtown activities but they are often close enough to be part of everyday activity space for downtown visitors.

A less common and perhaps more controversial element involves reindustrialization or at least the preservation of valued and sometimes picturesque blue-collar activities. In San Diego, the tuna fleet and associated tuna canneries and ship repair activities add a certain sense of place to the downtown waterfront. In Cleveland, Lake Erie freighters still ply the Cuyahoga River and port employment has actually gone up in recent years. In Columbus, a huge Ross Laboratory occupies a corner of the downtown near the inner belt and represents an attempt to get factory and warehouse activity back into the neighborhood. Portland has created a special warehouse refuge zone that is protected from higher-paying downtown land uses. In many downtowns, established support districts are being integrated into the downtown but are not necessarily changing it dramatically.

Downtown As an Event Center: The Diminished Perception of "Urban" Problems

For years it has been common for downtowns to be described as congested, polluted, and dangerous places that should be avoided by those who do not need to be there. The suburbs, on the other hand, were usually described as spacious, green, and quiet. During the past decade, much has changed and the charges once leveled almost exclusively at central cities are now commonly used to critique suburban areas as well.

Some of the changes reflect general trends in society. Crime is down almost everywhere, at least for a while, but the change in perception is most evident downtown. After decades of jokes about muggers and murders, Midtown Manhattan has become a family entertainment zone filled with Disney stores and arcades. Art galleries and T-shirt shops have largely replaced the sleazy bars of the San Francisco waterfront, and a shopping mall occupies much of the riverfront in New Orleans. The situation is much the same in all of the cities examined here.

As low-income neighborhoods and slum housing have been either cleared away or hidden behind megastructure barriers, the image of the downtown as poor and bedraggled has decreased. In a few cases, particularly in St. Louis, notorious close-in public housing projects have been cleared as uninhabitable. Part of the new perception may involve a lessening of racism coupled with the voluntary migration of people who can afford it away from inner city ghettos

as housing options have improved. On the other hand, part of the change may be due to the continuing reluctance of middle-class visitors to mix with the downtrodden and the increasing number of enclosed attractions such as malls and science centers. For a variety of complex and sometimes contradictory reasons, most downtowns are now considered to be relatively safe, family attractions, at least for special occasions, as well as business centers dominated by adults.

At the same time, suburban outer belt locations are being viewed as increasingly difficult and unfriendly places for the gathering of large numbers of people. Malls are overcrowded and parking lots are vast and uninviting. Highways and frontage roads lacking sidewalks predominate, and there are few, if any, public spaces. Special events such as parades, political rallies, and swap meets can be held in suburban parking lots, but the settings are often bleak.

For the new downtowns, massive crowds are signs of success rather than problems to be alleviated. Huge events such as marathons, car rallies, first nights (New Year's), Fourth of July fireworks, Christmas parades, carnivals, art and food festivals, open-air concerts, and a wide variety of other events now take place downtown, often in parks and other civic spaces but sometimes on major streets. Mass transit, special buses, and huge parking garages help, but crowding and congestion are hard to avoid. It could be that many people have become used to downtown crowds by going to ball games or concerts and associate them as much with pleasure as discomfort. Brochure and postcard images of "great" downtowns increasingly show huge crowds gathered for special events. Crowds, therefore, are no surprise. Downtowns are fun, while suburban strips have become associated with the ordinary workaday world.

All of this means selling the idea of reasonably easy access to downtown as well as a clear visual image or mental map of where to go once you arrive. Skyscrapers, waterfront promenades, huge shopping malls, and performing arts centers, when properly arranged, help to channel suburban visitors into and through pathways that are considered comfortable and safe. By way of contrast, outer belt business centers and edge cities often lack imageability and are especially intimidating for those who have left the safety of an automobile. The existence of a clearly understood grid and consensus "main streets" helps to give most downtowns a clear and comfortable image. Meandering suburban streets and disconnected frontage roads and cul-de-sacs, on the other hand, can make people less willing to explore.

Family-oriented events have played an important role in changing the image of downtowns as here at the Jazz and Rib Festival in Columbus.

A Composite Model of Downtown Organization

In this chapter, I have both summarized the individual place characteristics of each of the downtowns and attempted to create some generalizations or schematic models of the most common spatial patterns. In spite of a century of homogenizing trends, American downtowns remain distinctive. Nevertheless, consistent tendencies and spatial arrangements are apparent in most of the cities. As a conclusion, I present the final composite generalization: the linear, Disneyland-like, megastructure-buffered, amenity-oriented, civic-centered, greater downtown special event venue model. The model contains elements of each of the sixteen case studies examined as well as random and selected features from many other downtowns, both bigger and smaller. I suggest that this is what American downtowns are about as we enter the new millennium.

The pieces may vary from city to city, but the linear organization, amenity-orientation, big-footprint buildings, and distinctive subdistricts characterize nearly all successful downtowns. Also, the most successful downtowns are those that have maintained and/or developed strong connections to upscale residential neighborhoods. If at least some segment of the elite still thinks of

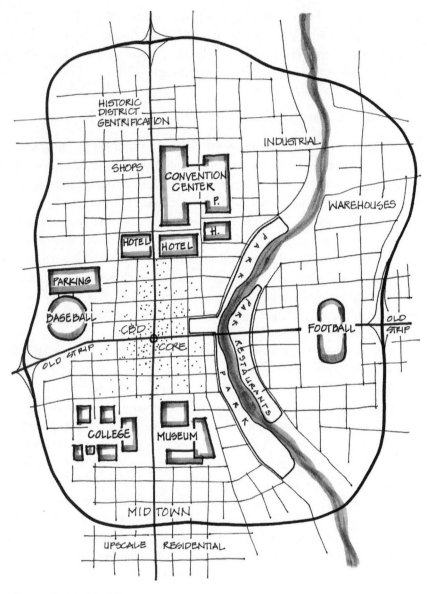

Composite Model of Downtown

downtown as its primary shopping, entertainment, and employment district, then support for new projects and maintenance is likely to be adequate. If downtown is viewed as a remote and deteriorating place used primarily by those who have no choice but to be there, the game is lost.

Where Do We Go from Here?

This book is meant to be a snapshot of selected downtowns around the year 2000. Obviously, there is much more that could have been included. For one thing, the number of downtowns examined could be increased dramatically so as to include underrepresented areas such as Florida and the Central South. On the other hand, it is doubtful that any completely new types of downtowns would emerge. Such efforts would probably expand our practical knowledge more than our conceptual frameworks. A second area for possible expansion would be to examine the future (especially long-range) plans for selected downtowns. I looked only at projects that were actually finished or under construction. But again, there is already a great deal of literature about big plans, including many that have not come to fruition.

Perhaps the biggest shortcoming in the book is the lack of emphasis on the actual people, the movers and shakers who made things happen and the organizational frameworks that were set up to facilitate growth and change. Obviously, most successful cities have had special downtown agencies overseeing redevelopment districts. These normally involve a number of dynamic planners, strong leadership in the public and private sector, lots of community involvement, and a certain amount of luck. I have looked into some of these things in a number of cities, but the stories are too long and complex to be interwoven with detailed descriptions of sixteen downtowns. Indeed, entire books, such as Thomas O'Connor's *Building a New Boston* have been written about the political intrigues involved in the revitalization of a single downtown.[3] I know from my experiences in San Diego that a similar book could be written about its Centre City Development Corporation (CCDC).

There is much work still to be done. I have tried to suggest, however, that it is useful to look at specifics—the details of spatial organization, architecture, and sense of place in real cities as we continue to discuss the changing roles of cities and downtowns in the America of the twenty-first century.

Notes

Introduction: The Downtown Imperative and the Need for Comparative Studies

1. Michael Sorkin, ed., *Variations on a Theme Park: The New American City and the End of Public Space* (New York: Noonday Press, 1992).
2. John Jakle and David Wilson, *Derelict Landscapes: The Wasting of America's Built Environment* (Savage, Md.: Rowman and Littlefield, 1992).
3. See for example, Shirley Laska and Daphne Spain, eds., *Back to the City: Issues in Neighborhood Renovation* (New York: Pergamon, 1980).
4. Raymond Murphy, *The Central Business District: A Study in Urban Geography* (Chicago: Aldine-Atherton, 1972).
5. Joel Garreau, *Edge City: Life on the New Frontier* (New York: Doubleday, 1991).
6. See for example, Peter Wolf, *Hot Towns: The Future of the Fastest Growing Communities in America* (New Brunswick, N.J.: Bantam Books, 1999).
7. See for example, Robert A. Caro, *The Power Broker: Robert Moses and the Fall of New York* (New York: Alfred A. Knopf, 1974).

CHAPTER 1: The American Downtown

1. See for example, Robert M. Fogelson, *Downtown: Its Rise and Fall, 1880–1950* (New Haven, Conn.: Yale University Press, 2001); Robert Liston, *Downtown: Our Challenging Urban Problems* (New York: Delacorte, 1965); Bernard Frieden and Lynnes Sagalyn, *Downtown, Inc.: How America Rebuilds Cities* (Cambridge, Mass.: MIT Press, 1990).
2. Arthur Herman, *The Idea of Decline in Western History* (New York: Free Press, 1997).
3. Stephanie Coontz, *The Way We Never Were: American Families and the Nostalgia Trap* (New York: Basic Books, 1992).
4. John Hannigan, *Fantasy City: Pleasure and Profit in the Postmodern Metropolis* (New York: Routledge, 1998).
5. Victor Gruen, *The Heart of Our Cities* (New York: Simon and Schuster, 1964).
6. Jane Jacobs, *The Death and Life of Great American Cities* (New York: Vintage Books, 1961).
7. See for example, Marshall B. Davidson, *Life in America*, vols. 1 and 2 (Boston: Houghton Mifflin, 1951).
8. See for example, Nathan Silver, *Lost New York* (New York: Schocken Books, 1967).
9. See for example, Jon Goss, "Disquiet on the Waterfront: Reflections on Nostalgia and Utopia in the Urban Archetype of Festival Marketplaces," *Urban Geography* 17

(1996): 221–47, and Allan Jacobs, "They're Locking the Doors to Downtown," *Urban Design International* 1 (1980): 25–27 and 38.

10. Ada Louise Huxtable, "Living with Fake and Learning to Like it," *New York Times,* 5 April 1997, p. 1.

11. Garreau, *Edge City.*

12. Carole Rifkind, *Main Street: The Face of Urban America* (New York: Harper and Row, 1977).

13. Edward W. Soja, *Thirdspace: Journeys to Los Angeles and Other Real-and-Imagined Places* (Cambridge, Mass.: Blackwell, 1996).

14. Raymond Murphy and James Vance, "Delimiting the Central Business District," *Economic Geography* 30 (1954): 189–222.

CHAPTER 2: The Evolution of the American Downtown, 1850–2000

1. Raymond Murphy, *The Central Business District* (Chicago: Aldine-Atherton, 1972).

2. James Vance, "Focus on Downtown," in *The Internal Structure of the City,* ed. Larry Bourne (New York: Oxford University Press, 1971), 112–20.

3. David Ward, *Cities and Immigrants: A Geography of Change in Nineteenth-Century America* (New York: Oxford University Press, 1971).

4. Mark Girouard, *Cities and People: A Social and Architectural History* (New Haven, Conn.: Yale University Press, 1985).

5. Kenneth Gibbs, *Business Architectural Imagery in America, 1870–1930* (Ann Arbor: University of Michigan Research Press, 1984).

6. Larry Ford, *Cities and Buildings: Skyscrapers, Skid Rows, and Suburbs* (Baltimore, Md.: Johns Hopkins University Press, 1994).

7. James Vance, "Human Mobility in the Shaping of Cities," in *Our Changing Cities,* ed. John Fraser Hart (Baltimore, Md.: Johns Hopkins University Press, 1990), 67–85.

8. Roberta Brandes Gratz and Norman Mintz, *Cities Back from the Edge: New Life for Downtown* (New York: John Wiley and Sons, 1998), 344.

9. Larry Ford, "Midtowns, Megastructures, and World Cities," *Geographical Review* 88 (October 1998): 528–47.

10. *The Urban Design Plan for San Francisco* (San Francisco: Department of City Planning, 1971).

11. Kevin Lynch, *The Image of the City* (Cambridge, Mass.: MIT Press, 1960).

12. Oscar Newman, *Defensible Space: Crime Prevention through Urban Design* (New York: Macmillan, 1973).

13. Tom Wolfe, *From Bauhaus to Our House* (New York: Pocket Books, 1982).

14. Ann Breen and Dick Rigby, *The New Waterfront: A Worldwide Urban Success Story* (New York: McGraw-Hill, 1996).

15. Michael Bernick and Robert Cervero, *Transit Villages of the Twenty-first Century* (New York: McGraw-Hill, 1997).

CHAPTER 3: The Downtown Stage

1. See for example, Larry Bourne and James Simmons, eds., *Systems of Cities: Readings on Structure, Growth, and Policy* (New York: Oxford University Press, 1978), and Larry Sawyers and William Tabb, eds., *Sunbelt/Snowbelt: Urban Development and Regional Restructuring* (New York: Oxford University Press, 1984).

2. Gruen, *The Heart of Our Cities.*

3. Stefan Lorant, *Pittsburgh: The Story of an American City* (Lenox, Mass.: Authors Edition, Inc., 1980).

4. Grady Clay, *Close-Up: How to Read the American City* (Chicago: University of Chicago Press, 1973).

5. Allan Jacobs, *Great Streets* (Cambridge, Mass.: MIT Press, 1993).

6. See for example, Anastasia Loukkaitou-Sideris and Tridib Banerjee, *Urban Design Downtown: Poetics and Politics of Form* (Berkeley: University of California Press, 1998), and Sorkin, *Variations on a Theme Park.*

7. William H. Wilson, *The City Beautiful Movement* (Baltimore, Md.: Johns Hopkins University Press, 1989).

8. Mike Davis, *City of Quartz* (New York: Vintage Books, 1992).

CHAPTER 4: Traditional Downtown Functions

1. John Costonis, *Space Adrift: Landmark Preservation and the Marketplace* (Urbana: University of Illinois Press, 1974).

2. Garreau, *Edge City*, p. 117.

3. Paul Groth, *Living Downtown: The History of Residential Hotels in the United States* (Berkeley: University of California Press, 1994).

CHAPTER 5: Downtown Expands

1. Tracy Newsome and Jonathan Comer, "Changing Intra-Urban Location Patterns of Major-League Sports Facilities," *Professional Geographer* 52, no. 1 (February 2000): 105–32.

2. Groth, *Living Downtown.*

3. Larry Ford, "Multiunit Housing in the American City," *Geographical Review* 76 (October 1986): 398.

4. Roman Cybriwsky, "Social Aspects of Neighborhood Change," *Annals of the Association of American Geographers* 68 (1978): 17–33.

5. Bernick and Cervero, *Transit Villages of the Twenty-first Century.*

6. Garreau, *Edge City.*

CHAPTER 6: Ranking Downtowns

1. J. R. Short, L. M. Benton, W. B. Luce, and J. Walton, "Reconstructing the Image of an Industrial City," *Annals of the Association of American Geographers* 83, no. 2 (June 1993): 207–24.

2. Whitney Seymour, ed., *Small Urban Spaces: The Philosophy, Design, Sociology, and Politics of Vest-Pocket Parks and Other Small Urban Spaces* (New York: New York University Press, 1969).

3. Thomas H. O'Connor, *Building a New Boston: Politics and Urban Renewal, 1950–1970* (Boston: Northeastern University Press, 1993).

Bibliography

AAA. 2000. Arizona, New Mexico; California, Nevada; Connecticut, Massachusetts, Rhode Island; Georgia, North Carolina, South Carolina; Illinois, Indiana, Ohio; Mid-Atlantic; New Jersey, Pennsylvania; North Central; Oregon, Washington Tourbook. Heathrow, Fla.: AAA Publishing.

Abbott, Carl. 1983. *Portland: Planning, Politics, and Growth in a Twentieth-Century City.* Lincoln: University of Nebraska Press.

————. 1993. "Five Downtown Strategies: Policy Discourse and Downtown Planning since 1945." In *Urban Public Policy: Historical Modes and Methods,* ed. Martin V. Melosi. University Park: Pennsylvania State University Press.

Abelson, E. S. 1996. "The City As a Playground: Culture, Conflict, and Race." *American Quarterly* 48: 523–29.

Adler, Jerry. 1995. "Theme Cities." *Newsweek,* 11 September: 68–70.

Anderson, Martin. 1964. *The Federal Bulldozer.* New York: McGraw-Hill.

Anderson, Stanford, ed. 1978. *On Streets.* Cambridge, Mass.: MIT Press.

Andrus, Phillip, et al. 1976. *Seattle.* Cambridge, Mass.: Ballinger.

Appleyard, Donald. 1981. *Livable Streets.* Berkeley: University of California Press.

Arizona State University Joint Urban Design Studio. 1995. *Downtown Phoenix Pedestrian Survey Update, Spring 1995.* Phoenix: Arizona State University.

Armstrong, Regina B. 1972. *The Office Industry: Patterns of Growth and Location.* New York: Regional Planning Association.

Arreger, Hans, and Otto Glaus. 1967. *High-rise Buildings and Urban Design.* New York: Praeger.

Arreola, Daniel. 1995. "Urban Ethnic Landscape Identity." *Geographical Review* 85, no. 4: 518–534.

Attoe, Wayne. 1981. *Skylines: Understanding and Molding Urban Silhouettes.* Chichester, U.K.: John Wiley and Sons.

Attoe, W., and D. Logan. 1989. *American Urban Architecture: Catalysts in the Design of Cities.* Berkeley: University of California Press.

Baade, Robert A. 1996. "Professional Sports as Catalyst for Metropolitan Development." *Journal of Urban Affairs* 18: 1–17.

Baade, Robert A., and R. F. Dye. 1990. "The Impact of Stadiums and Professional Sports on Metropolitan Area Development." *Growth and Change* 21: 1–14.

Bacon, Edmund N. 1974. *Design of Cities.* New York: Penguin Books.

Badcock, Blair. 1984. *Unfairly Structured Cities.* Oxford: Basil Blackwell.

Baerwald, Thomas. 1978. "The Emergence of a New Downtown." *Geographical Review* 68: 293–307.

Baltimore City Planning Department. 1985. *The Baltimore Harbor.* Baltimore, Md.: Baltimore City Planning Department.

————. 1999. *Plan Baltimore!* Baltimore, Md.: Baltimore City Planning Department.

Banerjee, Tridib, Genevieve Giuliano, Greg Hise, and David Sloane. 1996. "Invented and Reinvented Streets: Designing the New Shopping Experience." *Lusk Review* 2, no. 1: 18–30.

Banerjee, Tridib, and Michael Southworth. 1976. *Megastructures: Urban Features and the Recent Past.* New York: Harper and Row.

————. 1990. "Kevin Lynch: His Life and Works." In *City Sense and City Design: Writings and Projects of Kevin Lynch.* Cambridge, Mass.: MIT Press.

Barker, Emily. 1999. "Hotzones: The Best Cities in America for Starting and Growing a Business." *Inc.,* December: 67–85.

Barnes, Tom, and Dan Fitzpatrick. 1998. "Renaissance III." *Pittsburgh Post-Gazette,* 22 March: A-13.

Barnett, Jonathan. 1974. *Urban Design as Public Policy.* New York: McGraw-Hill.

————. 1982. *An Introduction to Urban Design.* New York: Harper and Row.

————. 1986. *The Elusive City: Five Centuries of Design, Ambition, and Miscalculation.* New York: Harper and Row.

————. 1995. *The Fractured Metropolis: Improving the Old City, Restoring the New City, Reshaping the Region.* New York: HarperCollins.

Barth, Gunther. 1980. *City People: The Rise of Modern City Culture in Nineteenth-Century America.* New York: Oxford University Press.

Bastian, Robert W. 1993. "Tall Office Buildings in Small American Cities 1923–1931." *Geografiska Annaler* 75: 31–39.

Bennett, Larry. 1990. *Fragments of Cities: The New American Downtowns and Neighborhoods.* Columbus: Ohio State University Press.

Bernick, Michael, and Robert Cervero. 1997. *Transit Villages of the Twenty-first Century.* New York: McGraw-Hill.

Berry, B. 1985. "Islands of Renewal in Seas of Decay." In *The New Urban Reality,* ed. P. E. Peterson. Washington, D.C.: Brookings Institution Press/Lincoln Institute of Land Policy.

Black, Thomas J. 1978. *The Changing Economic Role of Central Cities.* Washington, D.C.: Urban Land Institute.

Bluestone, Daniel M. 1991. *Constructing Chicago.* New Haven, Conn.: Yale University Press.

Bogue, Donald. 1963. *Skid Row in American Cities.* Chicago: Community and Family Study Center, University of Chicago.

Boston Redevelopment Authority. 1976. *Recycled Boston.* City of Boston.

Bosworth, Adrienne. 1990. "Creating a New West Bank." *Columbus Monthly,* August: 41–48.

Bourne, Larry S. 1996. "Reurbanization, Uneven Urban Development, and the Debate on New Urban Forms." *Urban Geography* 17: 690–713.

————, ed. 1971. *Internal Structure of the City: Readings on Space and Environment.* New York: Oxford University Press.

————, ed. 1982. *Internal Structure of the City: Readings on Urban Form, Growth, and Policy.* Oxford: Oxford University Press.

Bowden, Martyn J. 1971. "Downtown through Time: Delimitation, Expansion, and Internal Growth." *Economic Geography* 47: 121–35.

———. 1975. "Growth of the Central Business Districts in Large Cities." In *The New Urban History: Quantitative Explorations by American Historians,* ed. Leo F. Schnore. Princeton, N.J.: Princeton University Press.

Boyer, Christine. 1980. *Manhattan Manners.* New York: Rizzoli International Publications.

———. 1983. *Dreaming the Rational City: The Myth of American City Planning.* Cambridge, Mass.: MIT Press.

Bradbury, K. L., A. Downs, and K. A. Small. 1981. "Forty Theories of Urban Decline." *Urban Affairs Papers* 3: 13–20.

Breen, Ann, and Dick Rigby. 1985. *Caution: Working Waterfront: The Impact of Change on Marine Enterprises.* Washington, D.C.: Waterfront Press.

———. 1996. *The New Waterfront: A Worldwide Urban Success Story.* New York: Mc-Graw-Hill.

Bressi, Todd W., ed. 1993. *Planning and Zoning New York City: Yesterday, Today, and Tomorrow.* New Brunswick, N.J.: Center for Urban Policy Research.

Brill, Michael. 1989. "Transformation, Nostalgia, and Illusion in Public Life and Public Place." In *Public Places and Spaces,* ed. Irwin Altman and Ervin Zube. New York: Plenum Press.

Brooks, Jane S., and Alma H. Young. 1993. "Revitalizing the Central Business District in the Face of Decline." *Town Planning Review* 64: 251–71.

Bruegmann, Robert. 1982. "Two Post-Modernist Visions of Urban Design." *Landscape* 26: 31–37.

Bruttomesso, Rinio, ed. 1991. *Waterfronts: A New Urban Frontier.* Venice, Italy: Citta d'Acqua.

Bunce, O. B. 1954. "A Prophecy of Skyscrapers." *Landscape* 3: 26.

Buttenwieser, Ann L. 1987. *Manhattan Water-Bound: Planning and Developing Manhattan's Waterfront from the Seventeenth Century to the Present.* New York: New York University Press.

Carruth, Eleanor. 1969. "Manhattan's Office Building Binge." *Fortune,* October: 114.

Carter, Harold. 1975. *The Study of Urban Geography.* London: Edward Arnold.

CCDC. 1981. *Urban Design Program.* San Diego: Centre City Development Corporation.

Central Phoenix Committee. 1991. *Downtown Specific Plan.* Phoenix, Ariz.: Central Phoenix Committee.

Charlotte Chamber of Commerce. 1996. *Charlotte in Detail.* Charlotte, N.C.: Charlotte Chamber of Commerce.

Chicago, City of. 1986. *The Impact of a Major League Baseball Team on the Local Economy.* Chicago: Department of Economic Development.

Chidister, Mark. 1988. "Reconsidering the Piazza: City Life Requires New Design Models for Public Spaces." *Landscape Architecture* 78: 40–43.

———. 1989. "Public Places, Private Lives: Plazas and the Broader Public Landscape." *Places* 6: 32–37.

Chipello, Christopher J. 1999. "Death Knell for Domes." *San Diego Union Tribune,* 31 January: H-2 and H-4.

Christensen, Terry. 1982. "A Sort of Victory: Covent Garden Renewed." *Landscape* 26: 21–28.

Christian, C., and R. Harper. 1982. *Modern Metropolitan Systems.* Columbus, Ohio: Charles E. Merrill.

Chu, Henry. 1997. "Sidewalks Could Take New Path in Retailing." *Los Angeles Times,* 3 February: B-3 and B-1.

Chudacoff, H. 1975. *The Evolution of American Urban Society.* Englewood Cliffs, N.J.: Prentice-Hall.

Cianci, Vincent A., ed. 1999. *Good News: City of Providence.* Providence, R.I.: Vincent A. Cianci.

Clark, W. C., and J. L. Kingston. 1930. *The Skyscraper: A Study in the Economic Heights of Modern Office Buildings.* New York: American Institute of Steel.

Clay, Grady. 1973. *Close-Up: How to Read the American City.* Chicago: University of Chicago Press.

———. 1978. *Alleys: A Hidden Resource.* Louisville, Ky.: Louisville Design Center.

———. 1994. *Real Places: An Unconventional Guide to America's Generic Landscape.* Chicago: University of Chicago Press.

Cleveland City Planning Commission. 1998. *Downtown Plan, Cleveland.* Cleveland (Greater).

Cleveland Tomorrow. 1997a. *Civic Vision 2000 and Beyond.* Vols. 1 and 2. Cleveland: Cleveland Tomorrow.

Coffey, W. J., M. Polese, and R. Drolet. 1996. "Examining the Thesis of Central Business District Decline: Evidence from the Montreal Metropolitan Area." *Environment and Planning* A 28: 1795–1814.

Collins, Richard C., Elizabeth B. Waters, and A. Bruce Dotson. 1991. *America's Downtowns: Growth, Politics and Preservation.* Washington, D.C.: Preservation Press.

Columbus Chamber of Commerce (The Greater). 1989. "The 1988–'89 Guide to Central Ohio Offices and Industrial Parks." *Columbus Monthly Special Supplement.*

———. 1998. *Columbus: Positioned for Success.* Columbus C.E.O. May.

Columbus City Council. 1997. *Zoning Legislation.* Columbus: Columbus City Council.

———. 1994. *The Downtown South Plan.* Columbus: City of Columbus Development Department Planning Division.

———. 1995. *Columbus Historic District Maps.* Columbus: City of Columbus Development Department Planning Division.

Condon, George. 1967. *Cleveland: The Best-Kept Secret.* Garden City, N.Y.: Doubleday.

———. 1977. *Yesterday's Columbus: A Pictorial History of Ohio's Capital.* Miami, Fla.: E. A. Seemann Publishing.

Coontz, Stephanie. 1992. *The Way We Never Were: American Families and the Nostalgia Trap.* New York: Basic Books.

Cooper Marcus, Clare, and Carolyn Francis. 1990. *People Places: Design Guidelines for Urban Open Spaces.* New York: Van Nostrand Reinhold.

Costonis, John. 1974. *Space Adrift: Landmark Preservation and the Marketplace.* Urbana: University of Illinois Press.

Couvares, F. G. 1984. *The Remaking of Pittsburgh: Class and Culture in an Industrializing City, 1887–1919.* Albany: State University of New York Press.

Crilley, Darrell. 1993. "Megastructures and Urban Change: Aesthetics, Ideology, and Design." In *The Restless Urban Landscape,* ed. Paul Knox. Englewood Cliffs, N.J.: Prentice Hall.

Cybriwsky, Roman. 1978. "Social Aspects of Neighborhood Change." *Annals of the Association of American Geographers* 68: 17–33.

Cybriwsky, Roman, and John Western. 1982. "Revitalizing Downtowns: By Whom and for Whom?" *Geography and the Urban Environment* 5: 343–65.

Davidson, Marshall. 1951. *Life in America*. Vol. 2. New York: Houghton Mifflin.

Davis, Mike. 1991. "The Infinite Game: Redeveloping Downtown Los Angeles." In *Out of Site: A Social Criticism of Architecture*, ed. Diane Ghirardo. Seattle, Wash.: Bay Press.

———. 1992. *City of Quartz*. New York: Vintage Books.

Davis, S. 1997. *Spectacular Nature: Corporate Culture and the Sea World Experience*. Berkeley: University of California Press.

Dear, Michael J. 1986. "Postmodernism and Planning." *Environment and Planning* D: Society and Space 4: 367–84.

Dear, Michael J., and Jennifer Wolch. 1987. *Landscapes of Despair: From Deinstitutionalization to Homelessness*. Princeton, N.J.: Princeton University Press.

Deben, Léon, Sako Musterd, and Joan van Weesep. 1992. "Urban Revitalization and the Revival of Urban Culture." *Built Environment* 18: 85–89.

Demarest, M. 1981. "He Digs Downtown: For Master Planner James Rouse Urban Life Is a Festival." *Time*, 24 August: 36–42.

Denver, City and County of. 1982. *Denver Urban Design Sourcebook*. Denver: City and County of Denver Planning Office.

Diamonstein, Barbaralee. 1986. *Remaking America: New Uses, Old Places*. New York: Crown.

Dickinson, G. 1997. "Memories for Sale: Nostalgia and the Construction of Identity in Old Pasadena." *Quarterly Journal of Speech* 83: 1–27.

Dillon, David. 1996. "The Sage of the City, or How a Keen Observer Solves the Mysteries of Our Streets." *Preservation* September/October: 71–75.

Domosh, Mona. 1988. "The Symbolism of the Skyscraper: Case Studies of New York's First Tall Buildings." *Journal of Urban History* 14: 320–45.

———. 1990. "Shaping the Commercial City: Retail Districts in Nineteenth-Century New York and Boston." *Annals of the Association of American Geographers* 80: 268–84.

———. 1996. *Invented Cities*. New Haven, Conn.: Yale University Press.

Doughty, Martin, ed. 1986. *Building the Industrial City*. Leicester: Leicester University Press.

Dowall, David. 1986. "Planners and Office Overbuilding." *Journal of the American Planning Association* 52: 131–32.

Downs, Anthony. 1976. *Urban Problems and Prospects*. Chicago: Rand McNally.

———. 1997. "Alternative Futures for Inner-City Areas." *Lusk Review* 3: 23–40.

Downtown Columbus, Inc. 1988. *Downtown Columbus Strategic Plan: Task Force Recommendations*. Columbus, Ohio: Downtown Columbus, Inc.

———. 1992. *The Brewery District Plan*. Columbus, Ohio: Downtown Columbus, Inc. and City of Columbus Planning Division.

Downtown Denver Partnership, Inc. 1996. *Downtown Denver's Business Address*. Denver, Colo.: Downtown Denver Partnership, Inc. and Downtown Denver B.I.D.

Downtown Now! 1999. *St. Louis Downtown Development Action Plan.* St. Louis, Mo.: Downtown Now!

Downtown Phoenix Partnership. 1998a, b, and c. "Downtown Phoenix: The Place to Be!" *Downtown Update* 8 (spring, summer, and fall newsletters).

Downtown Seattle Association. 1996. "Downtown Seattle: A Work in Progress." *Washington CEO* November.

Drier, Peter, and Bruce Ehrlich. 1991. "Downtown Redevelopment and Urban Reform: The Politics of Boston's Linkage Policy." *Urban Affairs Quarterly* 26: 354–75.

Eckert, Kevin J. 1980. *The Unseen Elderly.* San Diego: Campanile Press.

Ellickson, Robert C. 1996. "Controlling Chronic Misconduct in City Spaces: Of Panhandlers, Skid Rows, and Public Space Zoning." *Yale Law Review* 105: 1165–1248.

Ellin, Nan. 1996. *Postmodern Urbanism.* Oxford: Blackwell Publishers, Inc.

———, ed. 1997. *The Architecture of Fear.* Princeton: Princeton Architectural Press.

Erenberg, L. 1991. "Impressions of Broadway Nightlife." In *Inventing Times Square,* ed. W. Taylor. New York: Russell Sage Foundation.

Euchner, Charles. 1993. *Playing the Field: Why Sports Teams Move and Cities Fight to Keep Them.* Baltimore, Md.: Johns Hopkins University Press.

Fainstein, Norman I., Susan Fainstein, and Alex Schwarz. 1989. "Economic Shifts and Land Use in the Global City: New York 1940–1987." In *Atop the Urban Hierarchy,* ed. Robert Beauregard. Totowa, N.J.: Rowman and Littlefield.

Fainstein, Susan S. 1991. "Promoting Economic Development: Urban Planning in the United States and Great Britain." *Journal of the American Planning Association* 57: 22–33.

———. 1994. *The City Builders: Property, Politics, and Planning in London and New York.* Oxford: Blackwell.

Faircloth, Anne. 1997. "North America's Most Improved Cities." *Fortune,* 24 November: 170–91.

Ferry, John W. 1960. *A History of the Department Store.* New York: Macmillan.

Firey, Walter. 1947. *Land Use in Central Boston.* Cambridge, Mass.: Harvard University Press.

Fisher, Lewis. 1996. *Saving San Antonio: The Precarious Preservation of a Heritage.* Lubbock: Texas Tech University Press.

Fitch, James Marston. 1990. *Historic Preservation: Curatorial Management of the Built World.* Charlottesville: University of Virginia Press.

Fjellman, S. M. 1992. *Vinyl Leaves: Walt Disney World and America.* Boulder, Colo.: Westview Press.

Flats Oxbow Association. 1995. *1995 Guide to the Flats, Cleveland.* Cleveland, Ohio: Flats Oxbow Association.

Fleissing, W. B. 1984. "The Yerba Buena Center." In *The City As Stage,* ed. K.W. Green. Washington, D.C.: Partners for Liveable Places.

Fleming, Ronald Lee, and Renata Von Tscharner. 1987. *Placemakers: Creating Public Art That Tells You Where You Are.* New York: Harcourt Brace Jovanovich.

Flynn, Raymond L. 1991. "Preface." In *Does America Need Cities? An Urban Strategy for National Prosperity,* ed. Joseph Persky, Elliot Selar, Wim Wiewel, and Walter Hook. Washington, D.C.: Economic Policy Institute.

Fogelson, Robert. 2001. *Downtown: Its Rise and Fall, 1880–1950.* New Haven, Conn.: Yale University Press.

Ford, Larry R. 1984a. "The Burden of the Past: Rethinking Historic Preservation." *Landscape* 28: 41–48.

———. 1984b. "Preserving Diversity: The Importance of Street-Level Doors." *California Geographer* 24: 1–20.

———. 1986. "Multiunit Housing in the American City." *Geographical Review* 76: 390–407.

———. 1992. "Reading the American Skyline." *Geographical Review* 82: 180–200.

———. 1994. *Cities and Buildings: Skyscrapers, Skid Rows, and Suburbs.* Baltimore, Md.: Johns Hopkins University Press.

———. 1998. "Midtowns, Megastructures, and World Cities." *Geographical Review* 88: 528–47.

———. 2000. *The Spaces between Buildings.* Baltimore, Md.: Johns Hopkins University Press.

Francaviglia, Richard V. 1977. "Main Street USA: The Creation of a Popular Image." *Landscape* 21: 18–23.

———. 1996. *Main Street Revisited: Time, Space, and Image Building in Small Town America.* Iowa City: University of Iowa Press.

Francis, Mark. 1988. "Changing Values for Public Spaces: Addressing User Needs Is Crucial to Success." *Landscape Architecture* 78: 54–59.

Franck, Karen A., and Lynn Paxson. 1989. "Women and Downtown Open Spaces." In *Public Places and Spaces,* ed. Irwin Altman and Ervin Zube. New York: Plenum Press.

French, Jere S. 1978. *Urban Space: A Brief History of the Urban Square.* Dubuque, Iowa: Kendall Hunt Publishing.

Friedberg, Anna. 1993. *Window Shopping: Cinema and the Postmodern.* Berkeley: University of California Press.

Frieden, Bernard. 1990. "Center City Transformed: Planners As Developers." *Journal of the American Planning Association* 56: 423–28.

Frieden, Bernard, and Lynne Sagalyn. 1990. *Downtown Inc.: How America Rebuilds Cities.* Cambridge, Mass.: MIT Press.

Friedrichs, Jurgen, and Allen C. Goodman. 1987. *The Changing Downtown: A Comparative Study of Baltimore and Hamburg.* New York: Walter de Gruyter.

Fujii, Tadashi, and Truman Asa Hartshorn. 1995. "The Changing Metropolitan Structure of Atlanta, Georgia: Locations of Functions and Regional Structure in a Multinucleated Urban Area." *Urban Geography* 16: 680–707.

Gad, Gunter, and Deryck Holdsworth. 1987a. "Corporate Capitalism and the Emergence of the High-Rise Office Building." *Urban Geography* 8: 212–30.

———. 1987b. "Looking Inside the Skyscraper: The Measurement of Building Size and Occupancy in Toronto Office Buildings, 1880–1950." *Urban History Review* 16: 176–89.

Garreau, Joel. 1991. *Edge City: Life on the New Frontier.* New York: Doubleday.

Garrett, Betty. 1980. *Columbus: America's Crossroads.* Tulsa, Okla.: Continental Heritage Press.

Gehl, Jan. 1987. *Life between Buildings: Using Public Space.* New York: Van Nostrand Reinhold Company.

Geist, Johann F. 1983. *Arcades: A History of a Building Type.* Cambridge, Mass.: MIT Press.

Gibbs, Kenneth. 1984. *Business Architectural Imagery in America, 1870–1930.* Ann Arbor, Mich.: UMI Research Press.

Girouard, Mark. 1985. *Cities and People: A Social and Architectural History.* New Haven, Conn.: Yale University Press.

Glazer, Nathan, and Mark Lilla, eds. 1987. *The Public Face of Architecture: Civil Culture and Public Spaces.* New York: Free Press.

Glennie, P., and Nigel Thrift. 1996. "Consumption, Shopping, and Gender." In *Retailing, Consumption, and Capital: Towards the New Retail Geography,* ed. N. Wrigley and M. Lowe. Harlow, Essex: Longman Group.

Goin, Peter. 1984. "Anonymous Places." *Landscape* 28: 24–29.

Goldberger, Paul. 1979. *The City Observed: New York.* New York: Vantage Books.

———. 1981. *The Skyscraper.* New York: Alfred A. Knopf.

———. 1989. "Seattle Tests Its Limits: Casting Votes on a Skyline's Future." *New York Times,* 16 May: B-1.

———. 1996. "The Rise of the Private City." In *Breaking Away: The Future of Cities,* ed. J. Vitullo Martin. New York: Twentieth Century Fund.

Goldfield, D. R. 1989. *Cotton Fields and Skyscrapers: Southern City and Region.* Baltimore, Md.: Johns Hopkins University Press.

Goldfield, Robert. 1998. "Portland's Affordable Housing Program Grows." *Puget Sound Business Journal,* 21–27 August: 17A.

Goldsteen, Joel B., and Cecil D. Elliott. 1994. *Designing America.* New York: Van Nostrand Reinhold.

Goode, James. 1988. *Best Addresses: A Century of Washington's Most Distinguished Apartment Houses.* Washington, D.C.: Smithsonian Institution Press.

Gordon, Jacques. 1985. *Horton Plaza: A Case Study of Private Development.* Cambridge, Mass.: MIT Center for Real Estate Development.

Gordon, Peter, and Harry W. Richardson. 1997. "The Density of Downtowns: Doom or Dazzle?" *Lusk Review* 3: 63–76.

Gornto, M. M. 1981. "Festivals Are Effective Promotional Tools." *Center City Report (International Downtown Executives Association),* June: 10–11.

Gosling, John R. 1998. "Patterns of Association." *Urban Land* 57: 42–47.

Goss, Jon. 1993. "The 'Magic of the Mall': An Analysis of Form, Function, and Meaning in the Contemporary Retail Built Environment." *Annals of the Association of American Geographers* 83, no. 1: 18–47.

———. 1996. "Disquiet on the Waterfront: Reflections on Nostalgia and Utopia in the Urban Archetype of Festival Marketplaces." *Urban Geography* 17: 221–47.

Gottdiener, Mark. 1997. *The Theming of America: Dreams, Visions, and Commercial Spaces.* Boulder, Colo.: Westview Press.

Gottman, Jean. 1966. "Why the Skyscraper?" *Geographical Review* 56: 190–212.

Gratz, Roberta Brandes, and Norman Mintz. 1998. *Cities Back from the Edge: New Life for Downtown.* New York: John Wiley and Sons.

Grogan, Paul S., and Tony Proscio. 2000. *Comeback Cities: A Blueprint for Urban Neighborhood Revival.* Boulder, Colo.: Westview Press.

Groth, Paul. 1994. *Living Downtown: The History of Residential Hotels in the United States.* Berkeley: University of California Press.

Grover, R., J. Weber, and R. A. Melcher. 1994. "The Entertainment Economy." *Business Week,* 14 March: 58–64.

Gruen, Victor. 1964. *The Heart of Our Cities.* New York: Simon and Schuster.

Hackney, Rod. 1990. *The Good, the Bad, and the Ugly: Crisis in Cities.* London: Fredrick Muller.

Haddix, Margaret Peterson. 1989. "Mr. Eli's Offer Led to Millions for City." *Indianapolis News*, 13 November: A-1 and A-8.

Hall, Peter. 1988. *Cities of Tomorrow.* Oxford: Basil Blackwell.

———. 1989. "The Turbulent Eighth Decade: Challenges to American City Planning." *Journal of the American Planning Association* 55: 275–333.

———. 1992. *Urban and Regional Planning.* London: Routledge.

Halpern, Kenneth. 1978. *Downtown USA: Urban Design in Nine American Cities.* New York: Whitney Library of Design.

Hanchett, Thomas W. 1998. *Sorting Out the New South City: Race, Class, and Urban Identity in Charlotte, 1875–1975.* Chapel Hill: University of North Carolina Press.

Hannigan, John. 1998. *Fantasy City: Pleasure and Profit in the Postmodern Metropolis.* New York: Routledge.

Harrison, J. 1997. "Museums and Touristic Expectations." *Annals of Tourism Research* 24: 23–40.

Hartman, Chester W. 1984. *The Transformation of San Francisco.* Totowa, N.J.: Rowman and Allanheld.

Hartshorn, Truman. 1980. *Interpreting the City: An Urban Geography.* New York: John Wiley and Sons.

Harvard Business School. 1984. "Cultural Revitalization in Six Cities." In *The City As a Stage*, ed. K.W. Green. Washington, D.C.: Partners for Liveable Places.

Harvey, David. "The Urban Face of Capitalism." In *Our Changing Cities*, ed. J. F. Hart. Baltimore, Md.: Johns Hopkins University Press.

Heffernan, Tony. 1995. "What's in the Games for Downtown Atlanta?" *Urban Land*, September: 43–44.

Heidel, Lynne. 1999. "Urban Canvas: Dreaming of What Could Be in Downtown San Diego." *San Diego Union Tribune*, 12 December: G-1 and G-5.

Herbert, David. 1972. *Urban Geography: A Social Perspective.* New York: Praeger Publishers.

Herman, Arthur. 1997. *The Idea of Decline in Western History.* New York: Free Press.

Hewison, R. 1987. *The Heritage Industry: Britain in a Climate of Decline.* London: Methuen.

Himmel, Kenneth A. 1998. "Entertainment-Enhanced Retail Fuels New Development." *Urban Land* 57: 42–46.

Histon, James. 1998. "Proposed Fifth and Forbes Retail and Entertainment District." *Pittsburgh Post-Gazette*, 29 July: State/Region.

Historic Seattle Preservation and Development Authority. 1975. *An Urban Resource Inventory for Seattle.* Seattle, Wash.: City of Seattle.

Holcomb, Briavel, and Robert Beauregard. 1981. *Revitalizing Cities.* Washington, D.C.: Association of American Geographers.

Holt, Nancy D. 1998. "Remaking Downtown Denver." *Urban Land*, April: 54–65.

Hough, Michael. 1984. *City Form and Natural Process.* London: Croom Helm.

———. 1990. *Out of Place: Restoring Identity to the Regional Landscape.* New Haven, Conn.: Yale University Press.

———. 1997. "How to Remake a City." *The Economist*, 31 May–6 June: 25–26.

Houstoun, Lawrence O., Jr. 1998. "Downtown Residential Conversions and Infill Construction Are Not Just Short-Term Phenomena." *Urban Land* 57: 35–41.

Hoyle, B. S., D. A. Pinder, and M. S. Husein, eds. 1988. *Revitalizing the Waterfront: International Dimensions of Dockland Redevelopment.* London: Belhaven Press.

Hughes, Kenneth H. 1998. "Transit on the Move." *Urban Land* 57: 9–11.

Hunker, Henry L. 2000. *Columbus, Ohio: A Personal History.* Columbus: Ohio State University Press.

Huxtable, Ada Louise. 1984. *The Tall Building Artistically Reconsidered: The Search for a Skyscraper Style.* New York: Pantheon Books.

———. 1997. "Living with Fake and Learning to Like It." *New York Times,* 5 April: 1.

Indianapolis Department of Metropolitan Development. 1991. *Indianapolis Regional Center Plan 1900–2010.* Indianapolis, Ind.: Department of Metropolitan Development Division of Planning.

Jacobs, Allan B. 1980. "They're Locking the Doors to Downtown." *Urban Design International* 1: 25–27 and 38.

———. 1985. *Looking at Cities.* Cambridge, Mass.: Harvard University Press.

———. 1993. *Great Streets.* Cambridge, Mass.: MIT Press.

Jacobs, Jane. 1958. "Downtown Is for People." In *The Exploding Metropolis,* ed. W. H. Whyte Jr. Garden City, N.J.: Doubleday.

———. 1961. *The Death and Life of Great American Cities.* New York: Vintage Books.

Jakle, John. 1985. *The Tourist: Travel in Twentieth-Century North America.* Lincoln: University of Nebraska Press.

———. 1987. *The Visual Elements of Landscape.* Amherst: University of Massachusetts Press.

Jakle, John, and David Wilson. 1992. *Derelict Landscapes: The Wasting of America's Built Environment.* Savage, Md.: Rowman and Littlefield.

Jameson, Fredric. 1991. *Postmodernism, or, The Cultural Logic of Late Capitalism.* Durham, N.C.: Duke University Press.

Jencks, Charles. 1977. *The Language of Post-Modern Architecture.* New York: Rizzoli.

Johannesen, E. 1984. "Cleveland's Circle and Square." In *The City As a Stage,* ed. K.W. Green. Washington, D.C.: Partners for Liveable Places.

Johnson, Dirk. 1997. "Once Stolid and Big-Shouldered, Now a Cinderella on the Lake." *New York Times,* 15 July: A-8V.

Judd, Dennis R., and Susan S. Fainstein, eds. 1999. *The Tourist City.* New Haven, Conn.: Yale University Press.

Kain, Roger. 1981. *Planning for Conservation.* London: Mansell.

Karp, Walter. 1982. *The Center: A History and Guide to Rockefeller Center.* New York: American Heritage Publishing Co.

Keating, W. Dennis. 1986. "Linking Downtown Development to Broader Community Goals: An Analysis of Linkage Policy in Three Cities." *Journal of the American Planning Association* 52: 133–41.

Keating, W. Dennis, and Norman Krumholz. 1991. "Downtown Plans of the 1980s: The Case for More Equity in the 1990s." *Journal of the American Planning Association* 57, no. 2: 136–52.

King, Anthony. 1980. *Buildings and Society: Essays on the Social Development of the Built Environment.* Boston, Mass.: Routledge and Kegan Paul.

———. 1996. "Worlds in the City: Manhattan Transfer and the Ascendance of Spectacular Space." *Planning Perspectives* 11: 97–114.

———, ed. 1996. *Re-presenting the City: Ethnicity, Capital, and Culture in the Twenty-First-Century Metropolis.* New York: New York University Press.

Kingston, J. L., and W. C. Clark. 1930. *The Skyscraper: A Study in the Economic Height of Modern Office Buildings.* New York: American Institute of Steel Construction.

Knox, Paul L. 1987. "The Social Production of the Built Environment: Architects, Architecture, and the Post-Modern City." *Progress in Human Geography* 11: 354–77.

Konvitz, Josef W. 1978. *Cities and the Sea: Port City Planning in Early Modern Europe.* Baltimore, Md.: Johns Hopkins University Press.

Kostof, Spiro. 1987. *America by Design.* Oxford: Oxford University Press.

———. 1991. *The City Shaped: Urban Patterns and Meanings through History.* Boston: Bullfinch.

———. 1992. *The City Assembled: The Elements of Urban Form through History.* Boston: Little Brown.

Kowinski, W. S. 1985. *The Malling of America: An Inside Look at the Great Consumer Paradise.* New York: William Morrow.

Kreisman, Lawrence. 1985. *Historic Preservation in Seattle.* Seattle, Wash.: Historic Seattle Preservation and Development Authority.

Krier, Rob. 1979. *Urban Space.* New York: Rizzoli.

Kunstler, James Howard. 1993. *The Geography of Nowhere: The Rise and Decline of America's Man-made Landscape.* New York: Simon and Schuster.

Laska, Shirley, and Daphne Spain. 1980. *Back to the City: Issues in Neighborhood Renovation.* New York: Pergamon.

Lassar, Terry. 1989. *Carrots and Sticks: New Zoning Downtown.* Washington, D.C.: Urban Land Institute.

Latham, J. E., ed. 1976. *The Economic Benefits of Preserving Old Buildings.* Washington, D.C.: Preservation Press.

Law, C. M. 1988. *The Uncertain Future of the Urban Core.* London: Routledge.

———. 1992. "Urban Tourism and Its Contribution to Economic Regeneration." *Urban Studies* 29: 597–618.

———. 1993. *Urban Tourism: Attracting Visitors to Large Cities.* New York: Mansell Publishing.

Lawrence, J. C. 1992. "Geographical Space, Social Space, and the Realm of the Department Store." *Urban History* 19: 64–83.

Leinberger, Christopher B. 1997. "The Coming Revival of American Downtowns." *Lusk Review* 3 (fall): 53–56.

Levine, M.V. 1987. "Downtown Redevelopment as an Urban Growth Strategy: A Critical Appraisal of the Baltimore Renaissance." *Journal of Urban Affairs* 9: 103–23.

Levy, John M. 2000. *Contemporary Urban Planning.* Inglewood Cliffs, N.J.: Prentice-Hall.

Lewis, Pierce. 1975. "To Revive Urban Downtowns, Show Respect for the Spirit of the Place." *Smithsonian*, September, 33–41.

Ley, David. 1980. "Liberal Ideology and the Postindustrial City." *Annals of the Association of American Geographers* 70: 238–58.

———. 1996. *The New Middle Class and the Remaking of the Central City.* Oxford: Oxford University Press.

Liebs, Chester. 1985. *Main Street to Miracle Mile: American Roadside Architecture.* Boston, Mass.: Little, Brown.

Lipton, S. 1977. "Evidence of Central City Revival." *Journal of the American Institute of Planners* 43: 136–47.

Liston, Robert. 1965. *Downtown: Our Challenging Urban Problems.* New York: Delacorte.

Lockwood, Charles. 1976. *Manhattan Moves Uptown.* New York: Houghton Mifflin.

Logan, Gregg, and Todd Noell. 1999. "Midtown Mania." *Urban Land,* April: 42–47 and 91.

Longcore, Travis R. 1996. "Information Technology and Downtown Restructuring: The Case of New York City's Financial District." *Urban Geography* 17: 354–72.

Longstreth, Richard. 1987. *The Buildings on Main Street: A Guide to American Commercial Architecture.* Washington, D.C.: Preservation Press.

Lorant, Stefan. 1980. *Pittsburgh: The Story of an American City.* Lenox, Mass.: Kingsport Press.

Lord, Rich. 1999. "Rust Belt Cities Plan 24-Hour Downtowns." *San Diego Union Tribune,* 5 September: A-29.

Lottman, H. 1976. *How Cities Are Saved.* New York: Universe Books.

Loukaitou-Sideris, Anastasia, and Tridib Banerjee. 1998. *Urban Design Downtown: Poetics and Politics of Form.* Berkeley: University of California Press.

Lowe, Jeanne. 1968. *Cities in a Race with Time.* New York: Vantage Books.

Lowenstein, Louis. 1971. *Urban Studies.* New York: Free Press.

Lubove, R. 1996. *Twentieth-Century Pittsburgh: The Post-Steel Era.* Vol. 2. Pittsburgh, Pa.: University of Pittsburgh Press.

Lynch, Kevin. 1960. *The Image of the City.* Cambridge, Mass.: MIT Press.

———. 1972. *What Time Is This Place?* Cambridge, Mass.: MIT Press.

———. 1981. *A Theory of Good City Form.* Cambridge, Mass.: MIT Press.

McChesney, Jim. 1999. "Urban Eden or Sprawling Hell?" *Oregon Quarterly,* Summer: 19–23.

McCormick, J. T. 1996. "Bringing Retail Back to Downtown Norfolk." *Urban Land* 55: 10–11.

McNair, Jeffrey. 1999. "Retail Entertainment." *Urban Land* 58: 35–37 and 94–95.

McNulty, R. H., R. L. Penne, D. R. Jacobson, and Partners for Liveable Places. 1996. *The Return of the Liveable City: Learning from America's Best,* Washington, D.C.: Acropolis Books.

McSheehy, William. 1979. *Skid Row.* Boston, Mass.: Schenkman.

Macdonald, Michael C. D. 1984. *America's Cities: A Report on the Myth of Urban Renaissance.* New York: Simon and Schuster.

Macdonald, S. 1996. "Theorizing Museums: An Introduction." In *Theorizing Museums,* ed. S. Macdonald and G. Fyfe. Oxford: Blackwell Publishers.

Manners, Ian. 1974. "The Office in Metropolis: An Opportunity for Shaping Metropolitan America." *Economic Geography* 50: 93–109.

Meredith, Robyn. 1997. "Demand for Single-Family Homes Helps Fuel Inner-City Resurgence." *New York Times,* 5 July: I-1 and 8-V.

Miller, C. Poh, and R. Wheeler. 1990. *Cleveland: A Concise History, 1796–1990.* Bloomington: Indiana University Press.

Miller, Michael. 1981. *The Bon Marché: Bourgeois Culture and the Department Store, 1869–1920.* Princeton, N.J.: Princeton University Press.

Miller, Robert L. 1998. "Narrative Urban Landscapes." *Urban Land* 57: 62–66.

Miller, Ronald. 1982. *The Demolition of Skid Row*. Lexington, Mass.: Lexington Books.

Minneapolis City Planning Department. 1996. *Downtown 2010: Continuing the Vision into the Twenty-first Century*. Minneapolis, Minn.: City Planning Department.

Minneapolis Planning Department. 1997. *The Minneapolis Plan: A Workbook for Citizen Comment*. Minneapolis, Minn.: Minneapolis Planning Department.

Moe, Richard, and Carter Wilkie. 1997. *Changing Places: Rebuilding Community in the Age of Sprawl*. New York: Henry Holt and Company.

Molotch, Harvey. 1996. "L.A. As a Design Product: How Art Works in a Regional Economy." In *The City: Los Angeles and Urban Theory at the End of the Twentieth Century,* ed. A. J. Scott and E. W. Soja. Berkeley: University of California Press.

Moore, Arthur Cotton, et al. 1971. *Bright Breathing Edges of City Life: Planning for Amenity Benefits of Urban Water Resources*. Springfield, Va.: Office of Water Resources, Department of the Interior.

Moore, Charles W. 1994. *Water and Architecture*. New York: Harry N. Abrams.

Motte, Mark T., and Laurence A. Weil. 2000. "Of Railroads and Regime Shifts: Downtown Renewal in Providence, Rhode Island." *Cities* 17: 7–18.

Mount Vernon Cultural District. 1998. *Go Visit the Mount Vernon Cultural District*. Baltimore, Md.: Trahan, Burden, and Charles, Inc.

Mozingo, Louise. 1989. "Women and Downtown Open Spaces," *Places* 6: 42–47.

Mujica, Francisco. 1977. *History of the Skyscraper*. New York: De Capo Press.

Muller, Edward K. 2000. "Downtown Pittsburgh Renaissance and Renewal." In *A Geographic Perspective of Pittsburgh and the Alleghenies: From Precambrian to Post-Industrial,* ed. Kevin J. Patrick and Joseph L. Scarpaci Jr. Pittsburgh, Pa.: Association of American Geographers.

Murphy, Raymond. 1955. "Internal Structure of the CBD." *Economic Geography* 31: 21–46.

———. 1972. *The Central Business District*. Chicago: Aldine-Atherton.

Murphy, Raymond, and James Vance. 1954a. "Delimiting the Central Business District." *Economic Geography* 30: 189–222.

———. 1954b. "A Comparative Study of Nine Central Business Districts." *Economic Geography* 30: 301–36.

"Museums, Theme Parks, and Heritage Experiences." 1991. *Museum Management and Curatorship* 10: 351–58.

Nasaw, D. 1993. *Going Out: The Rise and Fall of Public Amusements*. New York: HarperCollins.

NationsBank. N.d. *Fact Sheet: Welcome to NationsBank Corporate Center*. Charlotte, N.C.: NationsBank.

Netzer, Dick. 1978. *The Subsidized Muse: Public Support for the Arts in the United States*. Cambridge: Cambridge University Press.

Newman, Oscar. 1973. *Defensible Space: Crime Prevention through Urban Design*. New York: Macmillan.

Newsome, Tracy, and Jonathan Comer. 2000. "Changing Intra-Urban Location Patterns of Major-League Sports Facilities." *Professional Geographer* 52, no. 1: 105–32.

Norquist, John O. 1998. *The Wealth of Cities: Revitalizing the Centers of American Life*. Reading, Mass.: Addison-Wesley.

O'Brien, T. 1996. "Themed Eateries Thriving on Crowds of Others, But How Long Will It Last." *Amusement Business,* 4 March: 2–3 and 38.

O'Connor, Thomas H. 1993. *Building a New Boston: Politics and Urban Renewal 1950– 1970*. Boston, Mass.: Northeastern University Press.

Oldenburg, Ray. 1989. *The Great Good Place*. New York: Paragon House.

Olsen, Donald. 1986. *The City As a Work of Art*. New Haven, Conn.: Yale University Press.

Olsen, Sherry. 1976. *Baltimore*. Cambridge, Mass.: Ballinger.

Pagano, M. A., and A. Bowman. 1995. *Cityscapes and Capital*. Baltimore, Md.: Johns Hopkins University Press.

Peiss, K. 1986. *Working Women and Leisure in Turn-of-the-Century New York*. Philadelphia, Pa.: Temple University Press.

Perez-Pena, R. 1997. "Economists Dispute Value of Spending for Stadiums." *New York Times*, 3 August: 29 and 31.

Perry, Tony. 1999. "What the Staples Center Could Do for L. A." *Los Angeles Times Magazine*, 10 October: 51–60.

Phoenix, City of. 1991. *A Policy Plan for the Phoenix Arts District*. Phoenix, Ariz.: City of Phoenix Planning Department and EDAW, Inc.

———. 1997. *Central City Village*. Phoenix, Ariz.: City of Phoenix Planning Department.

———. 1998. *Downtown Phoenix Project Status Report*. Phoenix, Ariz.: City of Phoenix Community and Economic Development Department.

———. 2000. *Downtown Phoenix at a Glance*. Phoenix, Ariz.: City of Phoenix Community and Economic Development Department.

Pinck, Dan. 1998. "The Ideal City." *Preservation* Jan./Feb.: 34–36 and 82–85.

Pittsburgh, City of. 1998. *The Pittsburgh Downtown Plan*. Pittsburgh, Pa.: City of Pittsburgh Department of City Planning.

Pittsburgh Chamber of Commerce (Greater). 1997. *Greater Pittsburgh Region Metroguide: 1997–98 Annual Edition*. Pittsburgh, Pa.: Carson Publishing Co., Greater Pittsburgh Chamber of Commerce, and Pittsburgh Regional Alliance.

Pittsburgh Downtown Partnership. 1997. *Downtown Pittsburgh: America's Golden Triangle*. Pittsburgh, Pa.: Pittsburgh Downtown Partnership.

Pittsburgh Downtown Plan. 1997. *Retail Sales Comparison*. Pittsburgh, Pa.: Pittsburgh Downtown Plan and ZHA, Inc.

Portland, City of. 1983. *Downtown Design Guidelines*. Portland, Oreg.: City of Portland Bureau of Planning.

———. 1988. *Central City Plan*. Portland, Oreg.: City of Portland Bureau of Planning.

———. 1990. *Central City Plan Fundamental Design Guidelines*. Portland, Oreg.: City of Portland Bureau of Planning.

———. 1996a. *Central City Plan District Zoning Map Packet*. Portland, Oreg.: City of Portland Bureau of Planning.

———. 1996b. *Downtown Community Association's Residential Plan*. Portland, Oreg.: City of Portland Bureau of Planning.

Portland Bureau of Planning. 1992. Central City Developer's Handbook. Portland, Oreg.: Portland Bureau of Planning and Portland Development Commission.

Portland Metropolitan Association of Building Owners and Managers. 1991. *1991 Metropolitan Office Guide*. Portland, Oreg.: Portland Metropolitan Association of Building Owners and Managers.

Portland Progress (Association for). 1995. *Strategic Plan: Policy, Advocacy, and Programs*

for Greater Downtown Portland, 1995–2000. Portland, Oreg.: Association for Portland Progress.

Postman, Neil. 1985. *Amusing Ourselves to Death.* New York: Viking.

Powell's Books. 1997. *Walking Map of Downtown Portland.* Portland, Oreg.: Powell's Books.

Pratt Institute School of Architecture. 1997. "Streets: Old Paradigm, New Investment." *Places* 11: Special Addition.

Providence Warwick Convention and Visitors Bureau. 1999. *1999 Visitors Guide.* Providence, R.I.: Providence Warwick Convention and Visitors Bureau.

Pryne, Eric. 1997. "In City Life Not Dying; Downtown Population Is up 25%." *Seattle Times,* 14 August: A-1 and A-23.

Pygman, James, and Richard Kately. 1985. *Tall Office Buildings in the United States.* Washington, D.C.: Urban Land Institute.

Raitz, Karl, and John Paul Jones. 1988. "The City Hotel as Landscape Artifact and Community Symbol." *Journal of Cultural Geography* 9: 17–36.

Rannells, John. 1956. *The Core of the City: A Pilot of Changing Land Use in Central Business Districts.* New York: Columbia University Press.

Redstone, Louis. 1976. *The New Downtowns: Rebuilding Business Districts.* New York: McGraw-Hill.

Regional Plan Association. 1969. *Urban Design Manhattan.* New York: Viking Press.

Relph, Edward. 1976. *Place and Placelessness.* London: Pion Ltd.

———. 1987. *The Modern Urban Landscape.* Baltimore, Md.: Johns Hopkins University Press.

Reps, John. 1965. *The Making of Urban America: A History of City Planning in the United States.* Princeton, N.J.: Princeton University Press.

Riess, S. A. 1989. *City Games: The Evolution of American Urban Society and the Rise of Sports.* Urbana: University of Illinois Press.

Rifkind, Carole. 1977. *Main Street: The Face of Urban America.* New York: Harper and Row.

Riis, Jacob. 1890. *How the Other Half Lives: Studies of the Tenements of New York.* New York: Charles Scribner's Sons.

Riley, Robert. 1980. "Speculations on the New American Landscapes." *Landscape* 24: 1–9.

Roberts, S. M., and R. H. Schein. 1993. "The Entrepreneurial City: Fabricating Urban Development in Syracuse, New York." *Professional Geographer* 45: 21–33.

Robertson, K. A. 1995. "Downtown Redevelopment Strategies in the United States: An End-of-the-Century Assessment." *Journal of the American Planning Association* 61: 429–37.

Robinett, J., and D. Camp. 1997. "Urban Entertainment Centers Signal an Emerging Trend." *Urban Land,* March (Europe supplement): 20–22.

Rogers, Lynne Joy. 1997. "Technology and the Urban Economy." *Lusk Review* 3: 83–86.

Rosentraub, M. S. 1997. *Major League Losers: The Real Cost of Sports and Who's Paying for It.* New York: Basic Books.

Rotenberg, Robert, and Gary McDonogh, eds. 1993. *The Cultural Meaning of Urban Space.* Westport, Conn.: Bergin and Garvey.

Rouse, James. 1981. "He Digs Downtown." *Time,* 24 August: 42–53.

———. 1984. "The Case for Vision." In *Rebuilding America's Cities: Roads to Recovery,*

ed. P. R. Porter and D. C. Sweet. New Brunswick, N.J.: Center for Urban Policy Research.

Rubin, Barbara. 1979. "Aesthetic Ideology and Urban Design." *Annals of the Association of American Geographers* 69: 339–61.

Rubin, M., R. J. Gorman, and M. H. Lawry. 1994. "Entertainment Returns to Gotham." *Urban Land* 53: 59–65.

Ruchelman, Leonard I. 1977. *The World Trade Center: Politics and Policies of Skyscraper Development.* Syracuse, N.Y.: Syracuse University Press.

Rudofsky, Bernard. 1969. *Streets for People: A Primer for Americans.* Garden City: Doubleday.

Rutheiser, Charles. 1996. *Imagineering Atlanta.* New York: Verso.

Rybczynski, Witold. 1993. "The New Downtowns." *Atlantic Monthly* 271: 98–106.

St. Louis, City of. 1998. *Downtown Development Action Plan: Phase II.* St. Louis, Mo.: EDWA, Inc. et al.

———. 1999. *Downtown Development Action Plan: Phase II-C.* St. Louis, Mo.: EDWA, Inc. et al.

St. Louis Regional Commerce and Growth Association. 1999. *Saint Louis: America's Center for Business.* St. Louis, Mo.: St. Louis Regional Commerce and Growth Association.

St. Louis Union Station. N.d. *Memories: Celebrating the History of St. Louis Union Station.* St. Louis, Mo.: St. Louis Union Station.

San Antonio Downtown Alliance. 1999a. *Downtown Apartments and Residential Land Use Map.* San Antonio, Tex.: DTA.

———. 1999b. *Downtown Development Projects: September 1999.* San Antonio, Tex.: DTA.

———. 1999c. *Downtown San Antonio Map.* San Antonio, Tex.: DTA.

San Diego Building Owners and Managers Association. 1991. *San Diego Office and Business Parks Guide.* San Diego, Calif.: San Diego Building Owners and Managers Association.

San Diego CCDC. 2000. *Downtown Today.* San Diego, Calif.: Centre City Development Corporation.

San Francisco. 1971. *Urban Design Plan for San Francisco.* San Francisco, Calif.: Department of City Planning.

Sawyers, Larry, and William Tabb, eds. 1984. *Sunbelt/Snowbelt: Urban Development and Regional Restructuring.* New York: Oxford University Press.

Schaffer, Daniel, ed. 1988. *Two Centuries of American Planning.* Baltimore, Md.: Johns Hopkins University Press.

Schmandt, Michael. 1995. "Postmodern Phoenix." *Geographical Review* 85, no. 3: 349–63.

Schmidt, Steve. 1998. "They're Mad about Bob." *San Diego Union Tribune,* 25 September: A-1 and A-19.

Schubert, Dirk, and Anthony Sutcliffe. 1996. "The 'Haussmannization' of London: The Planning and Construction of Kingsway-Aldwych, 1889–1935." *Planning Perspectives* 11: 115–144.

Schuyler, David. 1986. *The New Urban Landscape: The Redefinition of City Form in Nineteenth-Century America.* Baltimore, Md.: Johns Hopkins University Press.

Seattle, City of. 1993. *Design Review: Guidelines for Multifamily and Commercial Buildings.* Seattle, Wash.: City of Seattle and Makers Architecture and Urban Design.

————. 1994. *Land Use Code: Technical and Procedural.* Seattle, Wash.: City of Seattle Department of Construction and Land Use.

————. 1995. *The Downtown Plan.* Seattle, Wash.: City of Seattle Office of Management and Planning.

Sennet, Richard. 1969. "Middle Class Families and Urban Violence: The Experience of a Chicago Community in the Nineteenth Century." In *Nineteenth Century Cities: Essays in New Urban History,* ed. S. Thernstrom and R. Sennett. New Haven, Conn.: Yale University Press.

————. 1977. *The Fall of the Public Man.* New York: Alfred Knopf.

Severini, Lois. 1983. *The Architecture of Finance.* Ann Arbor, Mich.: UMI Research Press.

Seymour, Whitney, ed. 1969. *Small Urban Spaces: The Philosophy, Design, Sociology, and Politics of Vest-Pocket Parks and Other Small Urban Spaces:* New York: New York University Press.

Sharpe, William, and Leonard Wallock. 1987. *Visions of the Modern City.* Baltimore, Md.: Johns Hopkins University Press.

Short, John R. 1996. *The Urban Order: An Introduction to Cities, Culture, and Power.* Oxford: Blackwell.

Short, John R., L. M. Benton, W. B. Luce, and J. Walton. 1993. "Reconstructing the Image of an Industrial City." *Annals of the Association of American Geographers* 83, no. 2: 207–24.

Shultz, Earle, and Walter Simmons. 1959. *Offices in the Sky.* Indianapolis, Ind.: Bobbs-Merrill.

Siegel, Fred. 1991. "Reclaiming Our Public Spaces." In *Metropolis: Center and Symbol of Our Times,* ed. Philip Kasinitz. New York: New York University Press.

Sigafoos, R. A. 1962. "What Is Happening to the Downtown Business District?" *Western City Magazine,* June: 25–26.

Silver, Nathan. 1967. *Lost New York.* New York: Schocken Books.

Simpson, Charles. 1981. *Soho: The Artist in the City.* Chicago: University of Chicago Press.

Skolnik, Arthur. 1976. "A History of Pioneer Square." In *Economic Benefits of Preserving Old Buildings,* ed. J. Latham. Washington, D.C.: Preservation Press.

Smith, Neil, and Peter Williams. 1986. *Gentrification of the City.* Boston, Mass.: Allen and Unwin.

Soja, Edward W. 1996. *Thirdspace: Journeys to Los Angeles and Other Real-and-Imagined Places.* Oxford: Blackwell.

Sorkin, Michael, ed. 1992. *Variations on a Theme Park: The New American City and the End of Public Space.* New York: Noonday Press.

Southworth, Michael. 1989. "Theory and Practice of Contemporary Urban Design: A Review of Urban Design Plans in the United States." *Town Planning Review* 60: 369–402.

Southworth, Michael, and Eran Ben-Joseph. 1997. *Streets and the Shaping of Towns and Cities.* New York: McGraw-Hill.

Spain, Daphne. 1992. *Gendered Spaces.* Chapel Hill: University of North Carolina Press.

Stanislawski, Dan. 1946. "The Origin and Spread of the Grid-Patterned Town." *Geographical Review* 36:105–20.

Starret, Paul. 1939. *Changing the Skyline.* New York: McGraw-Hill.

Starret, William. 1928. *Skyscrapers and the Men Who Build Them.* New York: Scribner's and Sons.

Stern, J. D. 1995. "An Amusement Park Moves Downtown." *Urban Land* 54: 13–14.

Stern, Robert. 1986. *Pride of Place*. Boston, Mass.: Houghton-Mifflin.

Stewman, Shelby, and Joel A. Tarr. 1982. "Four Decades of Public-Private Partnerships in Pittsburgh." In *Public-Private Partnerships in American Cities: Seven Case Studies*, ed. Scott R. Fosler and Renee A. Berger. Lexington, Mass.: Lexington Books.

Sudjic, Deyan. 1992. *The 100 Mile City*. San Diego, Calif.: Harcourt Brace.

Sutton, Horace. 1978. "America Falls in Love with Its Cities—Again." *Saturday Review*, August: 16–21.

Taylor, W. R. 1991. *Inventing Times Square: Commerce and Culture at the Crossroads of the World*. New York: Russell Sage Foundation.

Teaford, Jon C. 1990. *The Rough Road to Renaissance*. Baltimore, Md.: Johns Hopkins University Press.

Thrift, Nigel. 1993. "An Urban Impasse." *Theory, Culture, and Society* 10: 229–38.

Toffey, William E. 1985. *The Urban Form of Center City*. Philadelphia, Pa.: Philadelphia City Planning Commission.

Tunnard, Christopher, and Henry Reed. 1956. *The American Skyline*. New York: New American Library.

Urban Land Institute. 1983. *Urban Waterfront Development*. Washington, D.C.: Urban Land Institute.

Urry, J. 1990. *The Tourist Gaze*. London: Sage.

Vance, James. 1971. "Focus on Downtown." In *The Internal Structure of the City*, ed. Larry S. Bourne. New York: Oxford University Press.

———. 1990. *The Continuing City: Urban Morphology in Western Civilization*. Baltimore, Md.: Johns Hopkins University Press.

Van Leeuwen, Thomas. 1986. *The Skyward Trend of Thought*. Cambridge, Mass.: MIT Press.

Venturi, Robert. 1966. *Complexity and Contradiction in Architecture*. New York: Museum of Modern Art.

Venturi, Robert, Denise Scott Brown, and Robert Izenour. 1988. *Learning from Las Vegas*. Cambridge, Mass.: MIT Press.

Vernez-Moudon, Anne, ed. 1987. *Public Streets for Public Use*. New York: Van Nostrand Reinhold.

Wagner, George R. 1995. "Gentrification and Community: Baltimore's Inner Harbor." *25th Urban Affairs Association Annual Meeting*, 4 May: presentation draft.

Waller, G. A. 1995. *Main Street Amusements: Movies and Commercial Entertainment in a Southern City, 1897–1930*. Washington, D.C.: Smithsonian Institution Press.

Ward, David. 1966. "The Industrial Revolution and the Emergence of Boston's Central Business District." *Economic Geography* 42: 152–71.

———. 1971. *Cities and Immigrants: A Geography of Change in Nineteenth-Century America*. New York: Oxford University Press.

Ward, David, and Oliver Zunz, eds. 1992. *The Landscape of Modernity: Essays on New York City, 1900–1940*. New York: Russell Sage Foundation.

Warford, B., and B. Holly. 1997. "The Rise and Fall of Cleveland." *Annals of American Academy of Political and Social Science* 551: 208–21.

Warner, Sam Bass. 1972. *The Urban Wilderness: A History of the American City*. New York: Harper and Row.

Warren, Stacy. 1994. "Disneyfication of the Metropolis: Popular Resistance in Seattle." *Journal of Urban Affairs* 16: 89–107.

Warson, Albert. 1998. "Toronto's Waterfront Revival." *Urban Land* 57: 54–59.

Webb, Michael. 1990. *The City Square.* New York: Whitney Library of Design.

Weber, Bruce. 1997. "Cities Are Fostering the Arts As a Way to Save Downtown." *New York Times*, 18 November: special report.

Weinberg, Nathan. 1978. *Preservation in American Towns and Cities.* Boulder, Colo.: Westview Press.

Weisman, Leslie Kanes. 1992. *Discrimination by Design: A Feminist Critique of the Man-Made Environment.* Urbana: University of Illinois Press.

Weiss, Sonia. 1998. "Centre City Living." *Urban Land* 57: 61–65.

Westfall, Caroll William. 1987. "From Home to Towers: A Century of Chicago's Best Hotels and Tall Apartment Buildings." In *Chicago Architecture, 1872–1922: Birth of a Metropolis*, ed. John Zukowsky. Chicago: Art Institute of Chicago.

White, Norval. 1987. *New York: A Physical History.* New York: Athenaeum.

Whyte, William H. 1980. *The Social Life of Small Urban Places.* Washington, D.C.: Conservation Foundation.

———. 1988. *City: Rediscovering the Center.* New York: Doubleday.

———. 1993. "A 3D CBD: How the 1916 Zoning Law Shaped Manhattan's Central Business Districts." In *Planning and Zoning New York City*, ed. Todd W. Bressi. New Brunswick, N.J.: Rutgers University Press.

Willis, Carol. 1986. "Zoning and Zeitgeist: The Skyscraper and the City in the 1920s." *Journal of the Society of Architectural Historians* 45: 47–59.

———. 1995. *Form Follows Finance: Skyscrapers and Skylines in New York and Chicago.* New York: Princeton Architectural Press.

Wilson, E. 1991. *The Sphinx and the City.* London: Virago.

Wilson, William H. 1989. *The City Beautiful Movement.* Baltimore, Md.: Johns Hopkins University Press.

Wolf, Peter. 1999. *Hot Towns: The Future of the Fastest Growing Communities in America.* New Brunswick, N.J.: Rutgers University Press.

Wolfe, Tom. 1982. *From Bauhaus to Our House.* New York: Pocket Books.

———. 1998. *A Man in Full.* New York: Bantam Books.

Wright, T., and R. Hutchison. 1997. "Socio-Spatial Reproduction, Marketing Culture, and the Built Environment." *Research in Urban Sociology* 4: 187–214.

Wylson, Anthony. 1986. *Aquatecture: Architecture and Water.* New York: Van Nostrand Reinhold.

Young, Dwight, 2000. "A Rescued Fragment of Downtown Denver's History Still Thrives As a Preservation Pioneer." *Preservation* May/June: 108.

Zelinsky, Wilbur. 1994. "Conventionland USA: The Geography of a Latterday Phenomenon." *Annals of the Association of American Geographers* 84, no. 1: 68–86.

Zukin, Sharon, 1982. *Loft Living: Culture and Capital in Urban Change.* Baltimore, Md.: Johns Hopkins University Press.

———. 1991. *Landscapes of Power.* Berkeley: University of California Press.

———. 1995. *The Cultures of Cities.* Oxford: Blackwell.

Zunker, Vernon G. 1983. *A Dream Come True: Robert Hugman and the San Antonio Riverwalk.* San Antonio, Tex.: V. G. Zunker.

Index

About the Author

Larry R. Ford was born in Enid, Oklahoma, in 1943, and raised in a variety of places, mostly Columbus, Ohio. He received his B.S. and M.A. in geography at Ohio State University and completed his Ph.D. in geography at the University of Oregon in 1970. He then joined the faculty of the Department of Geography at San Diego State University, where he became a full professor in 1977. For the past thirty years he has published widely in leading professional journals in the areas of urban preservation, comparative city structure, and the relationship between architectural traditions and urban morphology. His books include *Cities and Buildings: Skyscrapers, Skid Rows, and Suburbs* (1994), *The Spaces between Buildings* (2000) and, with E. Griffin, *Southern California Extended* (1992).

Related Books in the Series